GOLD

and

A Hideaway of Your Own

Second Edition

by

Wentworth Tellington

Distributed by Gem Guides Book Co.
315 Cloverleaf Drive, Ste. F
Baldwin Park, CA 91706

DT & JJ Publishing Company
PO Box 716 • Downieville CA 95936

DT & JJ Publishing Company
P.O. Box 716 • Downieville, CA 95936

Library of Congress Catalog Card Number 92-71846
ISBN: 0-9630062-1-5
First Edition June, 1991
Revised August, 1991
Second Edition 1993
Printed in the United States of America, 1993

This nugget from the Deep Moon Gold Mine looked like a dirty golf ball with a showing of gold on one side when it was found. The only hint: It felt very heavy. After four hours in a hydrofluoric acid bath it came out shining bright with an amazing display of fine, lace-like gold threads around a central solid mass.

Come, let us on the sea-shore stand
And wonder at a grain of sand;
And then into the meadow pass
And marvel at a blade of grass;
Or cast our vision high and far
And thrill with wonder at a star;
A host of stars - night's holy tent
Huge-glittering with wonderment.

Yours is the beauty that you see
In any words I sing;
The magic and the melody
'Tis you, dear friend, who bring.
Yea, by the glory and the gleam
The loveliness that lures
Your thought to starry heights of dream,
The poem's yours,

Robert Service

The wonder of feeling one with everything in the great lonesomeness of a far off place learning where you fit between a blade of grass and the stars is something very special . . . and a hideaway such as we propose can truly hold the prospect of this unique discovery . . . still.

I sincerely hope you find some real benefit in this little book.

June 1993
Downieville, California, USA

YOUR OPPORTUNITY

Although the examples in this book are taken largely from the State of California, you can find opportunities to locate that hideaway of your own in all of the western mountain states and Alaska. Following is a list of the Bureau of Land Management offices that keep the official records:

HEADQUARTERS
U.S. Dept. of Interior
18th and C Street, N.W.
Washington, D.C. 20240
(202) 343-9435

ALASKA
701 C Street
Anchorage, AK 99513
(907) 271-5555

ARIZONA
3707 North 7th Street
Phoenix, AZ 85011
(602) 241-5504

CALIFORNIA
2800 Cottage Way, E-2841
Sacramento, CA 95825-1889
(916) 978-4746

COLORADO
2850 Youngfield Street
Lakewood, CO 80215-7076
(303) 236-1700

EASTERN STATES OFFICE
(covers other states)
350 South Pickett Street
Alexandria, VA 22034
(703) 274-0190

IDAHO
3380 Americana Terrace
Boise, ID 83706
(208) 334-1711

MONTANA (also covers North and South Dakota)
Granite Tower, 222 N. 32 Street
Billings, MT 59107-6800
(406) 657-6561

NEVADA
850 Harvard Way
Reno, NV 89520
(702) 784-5311

NEW MEXICO (also covers Texas and Oklahoma)
South Federal Place
Santa Fe, NM 87504-1449
(505) 988-6316

OREGON (also covers Washington)
825 NE Multnomah Street
Portland, OR 97208
(503) 231-6274

UTAH
324 South State Street
Salt Lake City, UT 84111-2303
(801) 524-3146

WYOMING (also covers Nebraska and Kansas)
2515 Warren Avenue
Cheyenne, WY 82003
(307) 772-2111

G O L D and A Hideaway of Your Own

by

Went Tellington

Contents

BRITISH COLUMBIA **Nº 70865** F

FREE MINER'S CERTIFICATE

NOT TRANSFERABLE

THIS IS TO CERTIFY that W. J. Tellington of 405 Petroleum Bldg Edmonton Alta, has paid me the sum of Five dollars, and is entitled to all the rights and privileges of a Free Miner, from midnight on the * 31st day of May, 1955, until midnight on the thirty-first day of May, 1956.

Issued at Victoria, the 11 day of May, 1955

[signature]
Signature of officer issuing same.

* Insert here the date of the day immediately preceding the day on which certificate is taken out.

British Columbia & Yukon Chamber of Mines

Membership Certificate

THIS IS TO CERTIFY that

W. J. Tellington, Esq.,

is a member of the British Columbia & Yukon Chamber of Mines for the year 1955.

MEMBER'S SIGNATURE

Stan E. Woodside
MANAGER

IT'S A FACT...

All over the western United States there are hundreds of thousands of mine claims on federal land in various conditions of ownership.

Most of them are about 20 acres in size. They are out in the desert, in the mountains, and along some of America's most beautiful streams.

Many of them are owned and controlled by major mining companies, but there are still thousands and thousands in private hands, or abandoned, or in some kind of limbo.

You can find these properties -
often in fabulously beautiful places -

waiting to be picked up - for practically nothing.

You only have to know where and how to look -
- and have a little perseverance.

It's fun just looking around in the country for these wonderful spots, and you could surprise yourself with the "pot of gold" that you could find "at the end of this rainbow."

HERE IS HOW YOU DO IT.

GOOD LUCK . . . AND HAPPY HUNTING!

The author's grandchildren with their parents have scrambled up the steep mountainside to the spring and dam - source of fabulous piped-in drinking water at the Deep Moon Gold Mine.

Friends, after a fine afternoon picnic at the Deep Moon Gold Mine, watch a professional gold miner pan and sluice pay dirt in the Downie River.

1. *OUR HERITAGE and PROSPECTING for GOLD*

While the basic thrust of this book involves finding a mine claim which you can call your own, it behooves us to consider some fundamentals of our culture and the ancient art of prospecting before we focus on the strategy, tactics, and technique of the mission.

Remember the old riddle: Which comes first, the hen or the egg?

There wouldn't be any mine claims to find if there were not a giant mine industry supplying us with essentials we need every day. This most critical dimension in our civilization has developed and continues to grow with the discovery of a broad variety of metals, minerals, and vital rock materials. All these things which we take for granted had to be found by someone; and the notion of prospecting goes back thousands of years.

If you are interested in the possibility of owning your own mine claim somewhere in the mountain or desert regions of the western United States, you would probably agree that there is a yearning deep inside you . . . a strange call back to the wilderness. A feeling, shared with the explorers in history that is hard to explain, but simply undeniable, expressed in the poem on the next page by George E. Winkler:

Creed For Prospectors

To love the touch of beauty on the rough stern hill,
The wild things of the thicket and the glade,
The furtive creep of Dawn and Eve's mesmeric still,
And mountain tarns of amethyst and jade.

To meet fate steadily, unshaken by each threat,
Prepared to do him battle at the bell;
To rally from his blows as one unvanquished yet
And ready to renew the bout in hell.

To keep the body active and the mind aware;
To follow Truth wherever she may lead;
To love the just and merciful and greed forswear,
And risk God damning you, for your creed.

Still, we must not forget the Yin and Yang of all this: One person's solitude is strangely and possibly eternally linked to the other's wish for the wonders of togetherness that concentrations of people can provide. The symphony you hear in a big city is not the symphony you hear by yourself in the wilderness. Quite obviously we need both.

In these last years of the Twentieth Century, finding that solitary place begins to look like an impossibility for many reasons. Driving in the country you may be trespassing on private property if you try to leave the public right of way. Even when you arrive in the vast public area we call the National Forest, you find signs saying: Do not leave the road. Camp only in designated areas. Wilderness Area - Keep Out.

Where for years we could escape to the country and pitch a tent by a stream or in some beautiful, solitary spot, we now find ourselves driving through a gate after paying an average of $14 to pitch a tent for one night. We park in the designated place, closer to other campers than we are to our neighbors in the city, and there is an arm's length of rules to obey . . . usually for very good reasons.

So what is a modern prospector going to do?

First — Recognize the need for preserving our environment for future generations, and resolve to behave in a way which complies with this fundamental moral obligation.

Second — Inform yourself, so you know what you can do and how to do it if you want to explore the back country and enjoy it.

* * * * *

The majority of people today would be afraid to go more than a few feet off a highway into a wilderness area for fear of animals or other threats to their safety, real or imagined. Yes, we have millions of hunters and fishermen who do go into the back country every year, but even they ordinarily return at night to a cabin or motel room. And the thought of being truly alone in the dark, a common experience among our forebears, with no street light where you can honestly enjoy the stars, is rare indeed.

Because of these simple facts of life, you can count on finding the wilderness as lonesome as you wish anytime you elect to go more than a hundred yards off the road, and especially in a spot where there is no path.

When it comes to looking for gold, there are many books devoted to the tactics and technique involved in exploring and recovering the elusive stuff. We have listed several in the bibliography at the back of this book, but one great text stands out:

Handbook for Prospectors and Operators of Small Mines by M.W. von Bernewitz, revised by Harry C. Chelson, McGraw Hill Book Co.

Out of print, this book is well worth having if you can find a copy. It represents the finest of the old knowledge. Since its publication, much has been discovered, and newer books have come along which keep us up to date; but the principles of exploration and prospecting have remained the same for centuries.

You will find it money well spent to acquire a good book on the geology of the earth. Do not settle for the kind of book commonly available in an American high school on earth science. These simplified versions, for the most part, suggest the fundamental object of a text at that level is as much focused on inculcating an acceptable political view of life on earth as imparting scientific information; and as a consequence there is little of practical value to a potential prospector for gold. You need a college level text of undisputed authority, such as:

Physical Geology, by Brian J. Skinner, et. al., published by John Wiley & Son, New York

This book, used at Yale University, is not only authoritative, but it is easy and enjoyable to read. We are not especially touting this book. There are at least a dozen other titles you might wish to purchase. Just be sure, if you are going to spend upwards of fifty dollars, that you buy good, solid information -- not some shallow, political argument arrogating itself as science.

We might ask: How do these scientists know so much anyway? They haven't been out to the stars. They haven't been to the center of the earth. They were not around millions of years ago. Most of them have never tried to actually find a gold mine. So how come?

In the upper echelons of academe, bright people are not only developing new stores of information, they are constantly improving the capacity to think through intellectual challenge and study. For example, we do not need to count hairs to prove beyond a doubt that there are at least two people walking around with exactly the same number of hairs on their heads.

It goes like this:

1. You can see hairs; and your ability to identify objects visually is limited to about 1/250th of an inch. Therefore, we have to allow that a hair is at least 1/250th inch in diameter, or we simply could not identify it.

2. We know there are several billion people on earth; and we have never seen a man with a head so large that it is three feet in diameter.

3. A bit of math will demonstrate that there is no room for a billion hairs 1/250th inch in diameter on a head three feet through. therefore, if you started counting people, saying to the first one, "You have one hair," (regardless of how many hairs the individual has); to the second, two, and so on, until you meet that one billionth person, you have to start over again because you have already established that no one can have a billion hairs.

4. At that point, logic leaves us no alternative but to conclude that somewhere along the way, we encountered a person who had the same number of hairs as another. The Greeks and Romans had a word for it: Q.E.D. Quod erat demonstrandum (Latin) which was to be demonstrated.

There is always the possibility that even a smart scientist could "shingle his roof into the fog" as we used to say in New England of the character who started the job, nailing with great zeal through a foggy morning until the sun burned clear, and he found himself precariously perched on nothing but shingles -- far beyond the roof he started on.

We would be foolish to ignore the wonderful achievements of today's scientists; but we cannot afford to allow ourselves to slip into the tempting abyss of big words here, and "shingle our roof into the fog."

There are two magic words: patterns and properties.

A property, as used here, is defined as the essential or distinctive attribute of something.

A pattern, as used here, is defined as any combination of materials or set of characteristics distinctive to a concentration of any given substance.

PROPERTIES The fundamental properties of gold are:

1. It is pale to deep yellow (depending on impurities) and appears the same from every angle.

2. It is amorphous, and relatively very soft. It can be scratched with a knife or hammered flat and very thin.

3. It will melt at a relatively low temperature and combines easily with other metals to make it wear better for use as jewelry.

Look at the nice little coffee can, you say. My gold mining friends, John and Bev Sieler, are having a good time with these "honest to God" nuggets demonstrating the two most profound properties;

4. It's very heavy, about twenty times as heavy as water. A one pound can like this will weigh about fifty pounds. No wonder John is holding both hands against his chest!

5. A can of gold this size is worth about a quarter of a million dollars in today's marketplace. No wonder Bev is grinning like a Cheshire cat!

There are other more sophisticated properties, but these are enough for our purpose as prospectors. Before proceeding on to the patterns, it is useful to note that gold is almost impossible to destroy. Old shipwrecks may ultimately give up to the deep, but any gold treasure is still there.

An authoritative writer has speculated that: "The death mask of a pharaoh eons ago . . . might have gone into the melting pot of a tomb robber to appear again as coins in the court of King Solomon, only to be remelted to become the drinking cup of a Roman . . . to appear in the crown of a king two thousand years from now, if kings are still around . . . after it has been in your ring. Gold is forever." Still, while hydrochloric acid (HCl) and nitric acid (HNO_3) have no effect on gold, it will dissolve in a combination of 3 parts HCL and 1 part HNO_3 called aqua regia (royal water).

Fool's gold, most commonly iron pyrite (FeS_2), is often mistaken for gold. It has a brassy yellow metallic luster and is too hard to scratch with a knife. It is often found near gold in a material lighter in both weight and color than the average rock. This gangue or matrix is called quartz (an old Czech word meaning hard). This observation leads us to patterns.

As a general statement we may say that gold is more often found in or near light colored rocks than dark colored rocks. Study of a good geology book will reveal why this is true.

To understand the patterns of gold we have to learn as much as we can about where it comes from, how it arrives at the location where you can find it, and what the characteristics of that location may be. This, as you know or soon will learn, is a very complicated subject. The good book on geology which I recommend will go into the principles of such vast subjects as how the earth was formed. Now, that is a running start if I ever heard of one. But as the old joke goes: You cannot undertake vast projects with half vast ideas.

Scientists make observations. They draw tentative conclusions. They attempt to duplicate their observations or find corroborative evidence which will fit some preestablished logic. Over decades, new ideas develop and replace older ones as these individual scientists hold meetings, write books, argue, and joust for preeminence in the cloistered world of their peers. Any time you stop the kaleidoscope you will hear a story which is presented as "the way it is." Experienced old guys with knarred hands in beer halls and coffee shops at the peripheries of civilization will also tell you where the gold is. When the chips are down, you are on your own.

PATTERNS   Where is this precious stuff?

It had to come from somewhere. There is no evidence that it dropped out of the sky, but there is ample evidence that it came up from deep inside the earth. So we accept that.

The accumulated wisdom of thousands of years in looking for gold and digging it up plus the organized professional analysis of the earth scene as interpreted by geologists is summarized in the simple diagram presented here for your thoughtful study. This represents the notion of an earth-surface building force, sometimes called the constructive force, which manifests itself in materials presumed to be hot enough to be relatively fluid, rising from somewhere below the surface up into or through the cooler materials already in place at the surface.

It would seem logical that where this stuff is so hot it could in theory burn its way through most anything, it might likely follow paths of weakness or least resistance such as cracks or soft zones in the crust. The geology book is full of ideas and words covering these possibilities, often drawn from actual observation of a variety of earth forms from big, old, multiple-cracked granite mountains to hot springs with complex forms of liquid coming out into strange, steaming pools. Being human, it is virtually impossible for these "experts" to resist the temptation to make up a story to fit what they find and examine. I am reminded of the old radio show where a comic character named Baron Munchausen asks the question: "Vas you dere, Charlie?" *For your amusement, let me say there really was such a person in the time of the American Revolution, Karl Frederich Hieronymus (1720-1797), Baron von Munchausen, a German soldier famous for his stories of impossible adventure as a cavalry officer in Russia. He is sometimes credited with the remark: Too soon old. Too late smart.* Most of us old prospectors would agree with that!

Geologists tell us the center of the earth is composed of a nickel-iron core. At the same time the scientists point out that gravitational forces tend to pull everything toward the center of the earth. There is a simple experiment involving small balls of various items with different densities or weights per unit volume. Placed in a container where they can be jarred steadily, the heavier ones will shift downward displacing the lighter ones and reach the bottom, where they cannot be unseated. This simple idea could lead one to wonder if there might not be a remarkable concentration of gold in that nickel-iron, or perhaps very close to the center because gold is so heavy. Some folks think so.

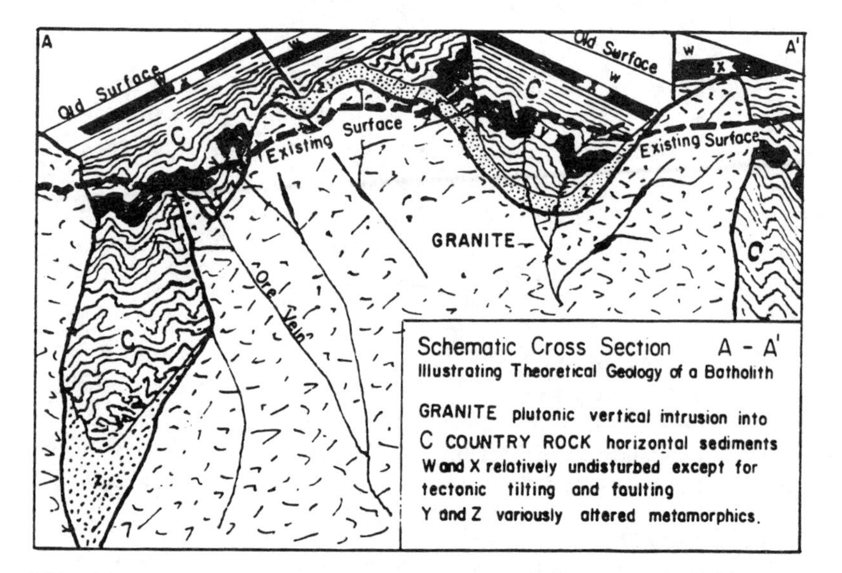

This drawing depicts a geologist's idea of an original set of more or less horizontal sedimentary rocks uplifted and displaced by a big bulge of hot granite and eroded down to the dashed line of a current *Existing Surface.*

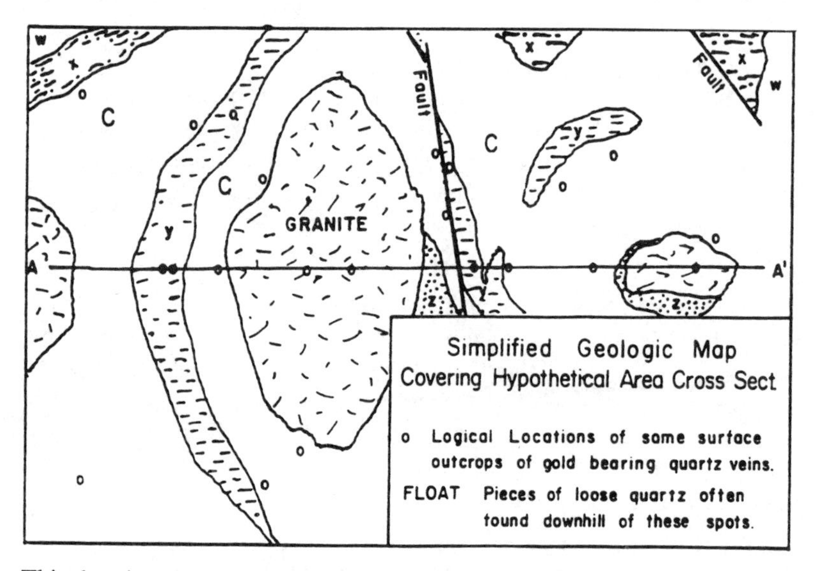

This drawing depicts a horizontal map of the above *Existing Surface* which a geologist would be able to make on the basis of ground observations and then imagine a subsurface configuration like the one in the cross-section. Scattered fragments of quartz would lead him to the gold bearing outcrops.

From those days centuries ago when Leonardo, wandering in the hills near his home in Italy, marvelled at the sight of rocks with precision images of sea shells indelibly stamped into their platy hardness far above the sea where such things would normally be expected to exist, we must admire the product of man's creative capacity to organize his observations of the evidence of events we know not of. Yet we must resist being expugnated by the forensic prolixity of self-gaudifying authority, and fall -- innocent victims of big words.

The processes of nature generally coming under the heading of geology seem to be eternally working to make changes in the landscape. Volcanoes erupt, earthquakes signal movements of large land masses and the oceans are constantly carving changes in the shorelines of the continents while rains fall, winds blow, and rivers wear new courses in their valleys.

Certain principles seem to govern the constructive and destructive activities as we see them on the surface of the earth. Gold, as we have noted, is heavy but can amalgamate with many lighter substances. It seems most commonly to "hitch a ride" from its natural location deep inside the earth due to its great weight and ride up in a matrix of very light stuff which becomes quartz.

The quartz containing blobs of soft gold is, itself, very light and brittle. When exposed to the destructive forces at the surface, chunks break away and begin to fall over the ground or in rivers. They hit other rocks or obstacles in place, and the quartz, being brittle, breaks. Since it is light it tends to roll down hill or flow downstream while the heavy blob of gold tends to stop right where it is in a low spot or behind a tree root and not move unless it is subject to more pounding or uprooting.

As rocks break up on their way from the mountain to the sea, the small and/or light ones travel the furthest and fastest while the big and/or heavy ones tend to stop short. Thus at all levels and in all grades of materials subject to these natural forces, we have a phenomenon involving foreset layering where the big, heavy things go the least distance and reside in relatively steep modes while the very light particles travel the greatest distances and reside in relatively horizontal modes.

We have common words which are associated with this phenomenon ranging from boulder through rock to pebble to gravel to sand to clay generally indicating size. Sand and clay can also suggest the kind of material where sand is associated with silica which is sharp and abrasive, while clay is associated with calcium and is soft and sticky.

High up on a mountainside where the rivers tumble away from their uppermost reaches, you can find large nuggets of gold near large boulders, and we have an old saying -- big rocks, big gold. Under the rocks where there may be accumulations of tiny particles, you will find gold possibly sealed by or in clay and attached almost like cement to the underside of big rocks -- well worth washing thoroughly to get the flakes and fines.

As you proceed down stream into more quiet water you can find mud and other fine particles of relatively light original rock with a specific gravity of between 2.0 and 4.0 with some exceptions. Here you can find very small particles of gold, still heavy, but now possibly trapped in other materials due to its natural affinities, and can even float in a moving current of water. Words like colloidal and micron begin to enter our vocabulary and we are instantly removed into a world of super scientific sophistication where big money is needed and big money can be made.

In this book we are confining ourselves to the world of an ordinary person hiking in the wilderness with ordinary funds, ordinary tools, and ordinary ambitions. Such an idea as having a metric ton of pure gold in 12.5 kilogram ingots approved by the Worshipful Company of Goldsmiths in the shadow of St. Dunstan, the patron saint of goldsmiths, delivered to some vault in Switzerland on some regular basis and bearing the ancient sign of the leopard's head is not what we are about here.

A few nice nuggets with "character" that we can give to a loved one on some special occasion, or at most a deposit worthy of modest "blocking out" for the purpose of floating a penny stock venture is really all we have in mind.

In our situation, we simply study enough to pick an interesting area where gold is known to have been found, and we start looking for a spot where we can dig to our heart's content.

We need some basic education; and it is fine if we can work with some experienced person who can teach us the fundamentals -- what to look for in place or in placer, and what to do about it. Then we simply join the ranks of the cheechakos, as Robert Service would tell you.

Cheechako is a word in Chinook jargon meaning newcomer, where t'shi means new and chako means to come. Common, informal language on the Trail of '98.

These pictures look pretty dull and uninteresting. But just imagine you and I are walking along a back country forest road and the bottom picture is of a rough cut-bank made by a bulldozer nearly uprooting that tree at the right. But wait. Those light colored rocks are pieces of quartz which we call "float" because they seem to float downhill on wet, slippery ground after a snow or rain. Where trees usually grow up more or less straight as they reach for the sun among other trees, this one seems to have a slight curve in the trunk. See how the roots seem to fall in a plane parallel to the sloping surface, then the trunk bends uphill as the main part of the tree assumes a vertical posture. This is called a "pistol grip" tree and reflects the slumping (slow sliding) of the mantle of thin dirt over the rock ledge underneath.

If we climb uphill (where the quartz must have fallen from) we come to a weak-looking, exposed ledge of rock (like weak slate or shale standing on edge). At the left the quartz appears intermittently in a more or less straight vertical line with tiny strings off to the left. Over to the right it seems to bend all around almost like branches of a tree.

We pound a few pieces loose and see shiny, flat surfaces near rusty looking spots, and then we see small bits with a different look -- not flat and shiny, but seeming to have a look of more depth, and appearing the same no matter how we hold the rock to the light.

You can dig back and mine for gold in place. This is a lode, and in our dreams it could lead us to the "Mother Lode." But we could say this is the source for pieces of gold broken free from the quartz down hill from here, among the roots of that tree. This is an excellent spot to pass a metal detector over and hear that tell-tale singing sound. Back in there you could find a piece of gold as big as your fingernail!

Down hill, in the stream which is bound to be not far below we can find concentrations of gold nuggets in any low spot, under any boulder, deep under the river bottom on top of a layer of clay or above the solid rock which we call "bed rock." The deepest part of the old river channel, on top of the solid rock, is called the "gut."

Where a river cuts into a bank, we call that the cut bank; and there is little chance for gold to hide in such a place. On the opposite side of the river (the inside bend) we usually find an area of gravel left over whenever the level of the water recedes leaving rocks of various sizes and pieces of wood or debris after a storm. That is the placer.

2. *THE BASIC IDEA . . .*

There may be other lands under other government controls in other places, but we are concerned in this book with those lands which lie west of the Mississippi River in the United States of America generally known as the public domain and administered by the U.S. Forest Service or the Bureau of Land Management.

The area covered is close to a half a billion acres.

Under federal mining laws resulting from statutes enacted in 1866, 1870, and 1872, a large body of case law and Supreme Court decisions, and Department of Interior rules and regulations enacted since those early days, we operate today generally under conditions prescribed by or similar to the BLM (Bureau of Land Management) "Surface Management Regulations."

There are four types of mine claims, of which we are primarily interested in only two. For the record, the four types are:

1. Lode claims (the most common) located on deposits of minerals encased in or surrounded by solid rock. These claims cannot exceed 1500 feet in length, along the center line of the deposit, and 300 feet on either side of the center line. They are up to 20 acres in size.

   You can see that these claims may have very odd shapes and odd orientations to the ordinary rectangular survey concepts since they generally follow naturally developed geomorphic forms.

2. Placer claims (the ones we are primarily interested in) located on deposits of loose, unconsolidated material or on consolidated sedimentary deposits lying at or near the surface.

   These claims are normally rectangular and conform to subdivisions of the federal section survey system unless conditions dictate irregular boundaries. They are generally 20 acres in size.

3. Mill site claims of up to 5 acres which may be located on non-mineral ground for the purpose of erecting a mill site or other processing plant or for a variety of other related purposes.

4. Tunnel site claims (rarely used) intended to protect the owner of a tunnel exploring blind lodes or veins, measured from the portal (entrance) up to 3,000 feet along the projected course of the tunnel and up to 1500 feet on either side of the center line of the tunnel.

With very rare exceptions, we can ignore types 3 and 4, and concentrate on lode and placer claims only.

All mining claims are initiated by posting a conspicuous notice on each claim showing the identity of the person locating the claim, the name of the claim, the date of location, and a brief description of the boundaries or dimensions. State laws may, in some instances, require more information.

Lode claims require monuments at the corners, as indicated in Chapter 5.

Placer claims only require one monument and a description of the claim in compliance with the federal land survey system, also discussed in more detail in Chapter 5.

Most of the mineral bearing or potentially mineral bearing land in the public domain is already located and properly filed with the appropriate authorities; so it is unlikely that you can simply go out into the forest and "stake a claim." It is possible, however.

You cannot just stake land because you like it. The law states that: only open, unappropriated, federal public domain containing valuable mineral deposits is open to location of mining claims.

The existence and location of lands like this is generally known among professional and amateur prospectors, but availability of specific lands for location can be confirmed if you go to the respective Federal Land Office, U.S. Bureau of Land Management, maintained for each separate western state.

Once a mine claim is located, it can be patented, converting it from an annually renewable lease to in fee simple ownership in the conventional real property sense. This requires specific compliance with detailed rules and regulations, and may involve a few years, but it can be and is done by individuals with and without the assistance of professional claim administrators or lawyers.

The basic idea for you is to acquaint yourself, as suggested in this book, with the "rules of the game," pick an area in the public domain which interests you, check in the BLM office in the state of your choice to get a feel for general availability, no matter who appears to own the land of your choice.

Using ordinary road maps in a general way and U.S.G.S. (United States Geological Survey) topographic maps for specific, detailed locations, you begin to reconnoiter the area of your choice. We explain the special topographic maps in Chapter 4.

There are pros and cons when we consider talking to local people in the area of your interest. Sooner or later you will probably be dealing with local people for one reason or another, but until you have a feel for a given area, it is probably just as well not to say too much about your objectives at first.

You will soon learn that you can find the land of your dreams from a variety of sources ranging from simply discovering some open land with minerals on it (rare) to discovering claims that are open to negotiation or relocation for one reason or another (not uncommon) or meeting a claim owner who is willing to sell for a price that sounds good to you (the easiest route).

Remember, claims can contain a great variety of minerals, and they must involve some truly acceptable mineral to be valid; but the most attractive claims usually involve precious metals such as silver or gold - and placer gold claims are, all in all, your best target because they often involve very attractive forest or desert land. Naturally, you will have to consider access. The most common access in federal lands is unpaved federal roads often maintained for fire fighting, lumbering, mining, or general management by the U.S.F.S. (United States Forest Service) or the B.L.M.

VARIOUS METHODS OF MINING IN THE EARLY DAYS

3. *MINERAL CLAIMS on FEDERAL LAND...*

As we have already suggested, there are opportunities to find a broad spectrum of mine claims wherever there is public domain and appropriate geology. In this manual we shall limit ourselves to essentially placer gold claims in California as an example of how you can find the "Hideaway of Your Own." The principles outlined can and do apply, however, from New Mexico to Alaska, and for any acceptable commercial mineral known to man. So do not think for a minute that you are limited to gold or California.

This writer has owned or enjoyed interests in large and small mineral properties from Alaska to Bolivia in South America , in British Columbia, Alberta and the Yukon Territory of Canada, and presently four beautiful gold bearing claims on the Downie River in Sierra County, California with some of the finest fishing and camping and pure enjoyment of the outdoors that you can imagine. We hope these simple facts may serve to convince you that:

YOU REALLY CAN FIND
THAT BEAUTIFUL HIDEAWAY OF YOUR OWN

After a person has discovered some evidence on the surface of the ground that there may be valuable minerals in the vicinity, and he or she is interested to pursue the discovery further, it is obvious that the person will want some kind of protection of his rights to the find if it does turn out to be valuable. That is the essential idea of a mineral claim and the reason for that choice of words.

The "Claim" is a notice of an understanding established by and between the locator and the United States government that the finder claims that valuable minerals may be in the area, and in compliance with applicable laws, the finder wants to prove his claim and subsequently extract the minerals for a profit. We must understand at the outset that the claim alone does not give the claim owner exclusive right to the ground except to mine it.

As long as you are actively mining it, however, you do have exclusive rights to the land. You can put up a fence and a locked gate if you wish, but when you are not operating, anyone at all can come across the land and fish or do anything else that may be acceptable to the public authority in the area - usually the U.S. Forest Service.

You can go to any U.S. Bureau of Land Management Office and get official pamphlets and bulletins telling you exactly how to locate a claim, and put up markers on the ground. They will also tell you exactly and officially what you can and cannot do, so we will not pursue this subject further here - preferring to be sure your information is absolutely correct.

Having made the appropriate moves on the ground, the next step is to fill out the Placer Mining Claim Location Notice specified by the BLM and have this document recorded in the county where it is located. This gives legal, constructive, public notice. The next move is to take this recorded document, keeping copies for yourself, to the BLM Office for filing, so the record goes into their computer.

Each year certain Assessment Work is required and/or fees to be paid and/or taxes. As a rule, when you are ready to file the annual notices required by the BLM, you must first have them recorded in the county; and they may not be willing to let you do this until they have some proof that all current taxes are paid, and the document carries official evidence to that effect.

There is such a thing as "claim jumping." It used to mean a case where some rough characters in the field literally came on the property and took it away at gun point. A slightly more refined act might be to simply put up stakes right over the original claim and then argue about it in a court of law or other appropriate forum. Today, it can be considerably more sophisticated, and you can lose your claim to some sharp individual who catches you at odds with the law or not in compliance with some technicality.

This is not a serious problem if you simply pay attention to business and make sure you do things on or before the dates and times specified by law. The two basic agencies to deal with are the BLM and the county in the state where your claims are located. If you have any questions, they have people who are glad to help you.

As long as you comply with the law and the applicable rules and regulations, there is no reason why you can not maintain your claims for years and years and turn them over to your heirs, successors, or assigns, as they say, in perpetuity.

It is possible that you could pick an area, go to it and look around; and then, having found what seems like a good spot, check it in the BLM Office by using their computer service. (They will help you do this.) Then, to your great delight, you might discover that it is "Open to location." But we wouldn't bet on it!

You will learn that just about every good looking spot in every state and county of the western United States appears to be already taken. But, do not be discouraged.

There are thousands of good claims in the hands of older people or absentee owners or others who will often, for a modest price, sell them to you. There are thousands of claims in the hands of miners who need money or want to move on to "richer looking pastures." That seems to be the very nature of many prospector types everywhere.

There are people who deal in these things, and will trade you a dead cat for a dead dog any day and twice on Sunday! You should be wary much of the time. On the other hand, we must say that there are many legitimate traders in mine claims who can be very helpful.

You may discover by a process of due diligence in the BLM and the local county office that a claim you would like is indeed in some kind of legal limbo. In these cases you have to be sure that the named owner of record is not protected by some statute of recovery wherein he has some grace period within which he may reinstate his ownership. Here is where your skills in pseudo-legal activities may prove useful. People follow these routes in many ways besides modern claim jumping by watching the legal notices in newspapers and going to various kinds of public auctions or other activities carried out by authorities to deal with unsettled accounts.

One of the simplest and most straightforward things to do is to go to an interesting area and look at the bulletin boards in the small towns. You will find the usual assortment of ads and notices from babysitters to mobile homes, and among them it is not uncommon to see a gold dredge for sale, or a pump, or a winch, or an old 4x4 vehicle, or a piece of real estate, or even a mine claim.

It could be that easy, if the price is right.

The thing you have to be sure about is that you are getting what you think you are, and it is worth what you think it is, and you are not buying some big hidden problem. But that's life.

Country bulletin boards like this one frequently announce gems of information for the claim hunter.

Wherever you go looking for this "dream hideaway" there will probably be people ready to offer you assistance. You know that!

A good way to learn about an area is to find out about the local papers. Buy a few issues. Read everything - the regular ads, letters to the Editor, public notices - the works.

If you run into someone you think you might do business with, and the scope of your deal is worth the trouble, just go to the Recorder's Office and look for that person's name in the Grantor/Grantee Records for the past few years. It requires only a few minutes, and is a very worthwhile activity.

Season Tickets $50
1991 Concert Series
Kentucky Mine Bowl

Mexican Dinner May 10
Downieville Hall 5:30
by Search & Rescue

Firemen's Follies
Sierra City hall
May 25th & 26th

The Mountain Messenger

California's Oldest Weekly Newspaper

VOL. 137 NO. 44 THURSDAY, MAY 2, 1991 DOWNIEVILLE, SIERRA COUNTY, CALIFORNIA ESTABLISHED 1853 SINGLE COPY 25 CENTS

Supes To D.C.? Unhappy About Land Swap

by Liz Fisher
Research by Tim Smith
SCTV CH-9 Sierra City

DOWNIEVILLE A trip to Washington, D.C. may be in store for two Sierra County Supervisors in the attempt to convince the U.S.Forest Service to move the Downieville District Ranger Station back to the county seat.

Chairman of the Board, Jerry McCaffrey and Vice Chairman, DonMcIntosh,hopeto meet with Congressional and Administrative representatives to discuss aspects of the situation which has become complicated with land exchange issues.

A proposed Land Exchange between the U.S.Forest Service and Alpine Meadows,Inc. of Tahoe has become a key point in the effort to convince the USFS to return their district office from Camptonville to Downieville.

According to USFS documents the land exchange will include two parcels, one of 105 acres at the base of Alpine Meadows Ski Resort, including a parking lot and commercial buildings, the other a 22 acre site near the intersection of the Alpine Meadows entrance road and Highway 89. These parcels of

Ken Werner, Land Exchange Officer for the USFS says the exchange is necessary as the Forest Service is not allowed to make cash purchases of land nor sell public lands for cash.

The negotiations to acquire title to the two private parcels, owned by Willis Nelson, will result in Alpine Meadows paying $1,755,000 to Nelson to acquire title to the properties in Camptonville and Nevada City. During a simultaneous escrow Alpine will then transfer title to the USFS and receive title to the 127 acres at the Alpine Meadows Ski Resort.

County representatives believe that USFS ownership of the Camptonville property would allow avoiding the issue of returning headquarters to the Downieville Ranger Station, property the USFS currently owns.

The USFS signed an agreement with Alpine Meadows in May of 1990 renewing the Special Use Permit for a 40 year period.

The Special Use Permit is for an area in excess of 640 acres," said Werner, "not just the 127 acres in the exchange proposal."

Werner said the 40 year period is a standard length of time and replaces a 25 year SUP issued in 1978.

come and various other factors," said Werner.

Werner felt the loss of fees due to the 122 acre exchange would be "minimal."

Realtors in the Tahoe area where Alpine Meadows is located said a 40 year Special Use Permit would have the effect of decreasing the land value substantially.

One Realtor said land in the Deer Park area, the 22 acre location, would be a "bargain at $150,000 an acre, the average roadfront property would be $200,000 to $300,000 per acre." According to USFS figures, the value of the lands average $13,819 per acre.

Werner responded to this statement by saying, "You have to be careful about what Realtors in the Tahoe area say, it could cause some misunderstanding."

Werner said the USFS does not buy or sell property on a cash basis, "we're only able to do equal value land exchanges and the value established on the Alpine Meadows property was by an independent real estage appraiser from Auburn."

McCaffrey and McIntosh hope to convince Congressional representatives to support the return of District Headquarters to

Members of the Giants, a winning team, are: Coach Mike Lozano, Summer Villarreal, Erin Sellards, Summer Svalberg, Amy Olson, Jenny Schofield, Celeste Villarreal, Jessi Black, Amber Baca, Melissa Floyd, Coach Pete Villarreal, Coach Brenda Black, Marie Brown, Autumn Brown, D.J. Hunter, Christy Lozano, Danielle Kalefel and Heather Best.

Local A's Drop Squeaker

DOWNIEVILLE-The Little League Minors (ages 8,9,10) lost a close game Tuesday in Portola after five full innings of play. The Downieville-Sierra City Team called the Athletics scored six runs

Downieville pitcher, Aaron Kirby pitched the full game and the entire team displayed good fielding and hitting for the first game of the season. The Portola Giants came back with five runs in the fifth in-

Rinker made good plays in the field and good base running from Ben Johnson, Berto Landa and others helped the team with the seven runs. The entire team displayed great sportsmanship and

You never know what you will find in a country paper like this.

Many promising deals have gone awry for lack of ordinary facts.

In most cases, the last thing you want to do is get a lawyer involved. In simple cases, experience teaches that he will scare people away who might ordinarily be willing to initiate a deal on the back of a napkin in the local cafe, or he will botch the deal completely; and in complicated cases, he will cost you more than the average mine claim could possibly be worth.

You have to count on doing your own leg work for the most part. However, if you take your time, you can learn in any given locality who the people are that you can count on in a general way. But there is no arguing that in mine deals, as a rule, there is no lack of prevarication. Keep your guard up. The gold nuggets grow heavier and brighter with every "telling"!

<u>It is commonly agreed that many a "good" mine has been spoiled with a pick</u>.

Most placer gold claims are real enough. You can, as a rule, find "some color." That is to say with careful panning of the right gravel in the right place you can find little pieces, "fines or flakes" big enough to be seen with the naked eye. And many claims have yielded real, first class, honest-to-goodness, jewelry quality nuggets.

We cannot leave this phase of the subject without at least mentioning the ancient and respectable art of "salting" a mine. This manual is much too short to cover anything other than the highlights, but suffice it to say there is many a soul leaning back on a bench outside the local grocery stores in about every mining town in the West with the skills of a magician when it comes to sleight of hand near a gold pan. Just two examples will do:

Beware the smoker with the cigarette dangling out of his mouth as he pans. An "ash" can fall innocently into the pan as he sloshes the gravel around. A little gold concealed in the cigarette falls in, and the ash floats away. This is a crude trick that most skilled salters would scoff at as being far beneath their dignity!

Before you ever arrive, the seller of the mine could blast a rock or a bed of sand with a light shotgun load containing a bit of the beautiful dust, so it looks as though it has been there forever!

Oh well, mine claims and gold deals wouldn't be any fun at all without some entertainment, would they?

There is another side to this subject, before we move along:

You could be on to a really good source of gold. In that case, you would go through the process of patenting the claim and converting it to regular ownership of real estate which would have very significant value in the mountains of the west where people generally agree that even "goat pasture" will bring a very fine price if you can get to it.

The fundamental notion in this situation is what is referred to as "the prudent person test" of discovery as laid down in *Castle v. Womble, 19 LD 455 (1894)* in which the Secretary of the Interior of the United States stated: "where minerals have been found and the evidence is of such a character that a person of ordinary prudence would be justified in the further expenditure of his labor and means, with a reasonable prospect of success, in developing a valuable mine, the requirements of the statutes have been met."

Many people will tell you that it is "practically impossible to patent a mine claim today," but let's look at the facts.

Even though there are about 1.5 million mine claims in the West, during the year 1981, for example, only 51 applications were filed for patent. During that period 40 patents were issued. It looks as though the only reason that few patents are issued is that very few applications are made in the first place.

People will also tell you "It takes forever to get a patent." Not so. This writer knows a person locally who did his own work - didn't even hire a lawyer, and got his mine patented without mining the land in less than two years. Ideally the whole process can be handled in a single year, but realistically you can figure on two to five years and a relatively reasonable cost if you are smart.

Patenting is an opportunity, beyond simply finding a neat place to camp and have a beer while your "old man" gets the fire going, and fires up the big trout you just caught, where you can turn that somewhat fragile lease, called a claim, into a full title! Sometimes it takes a woman's perseverance to see past the romance of a gold pan, and get to the nitty-gritties!

So much for the general subject of mine claims. Let's move on to consider some of the technicalities. We shall discuss the actual location and maintenance of claims in Chapter 5 . . . Locating Claims and Official Records.

FOLDING A MAP

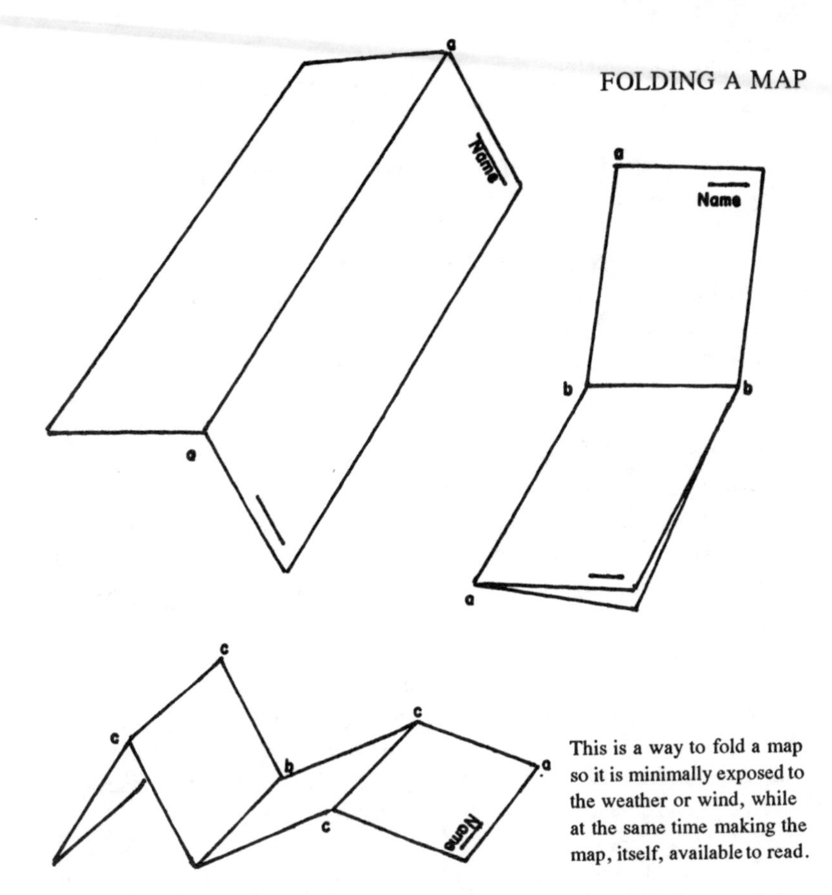

This is a way to fold a map so it is minimally exposed to the weather or wind, while at the same time making the map, itself, available to read.

This instruction is specifically identified with U.S.G.S. standard topographic quadrangles where the primary name is located at the top and bottom right side of the paper.

Step 1: Fold the map along a center line a-a so the map is visible on both sides of the fold, as shown in the diagram.

Step 2: Fold the map again along secondary centerline b-b so the names at top and bottom face each other, and are not immediately visible.

Step 3: Finally fold the map halves back on each other as shown in the diagram, so the names face out again.

Now you see the area of the map which is exposed at any time will ordinarily be only 1/8 the size of the map as a whole, and at the most only 1/4, so it is easier to protect from the elements.

N.B: For historical buffs, this technique was first introduced in the French Army under Napoleon when he ordered the first maps of this kind published for Europe during the first years of the 19th Century.

4. *A WORD ABOUT MAPS . . .*

Since maps are critical to our mission to find gold and that hideaway, we will spend a short time discussing the kinds of maps that can be useful and the places where you find them.

We assume that you are familiar with the various types of atlases or other common maps in daily use for the purpose of finding places or determining how to travel from one place to another: Rand McNally road atlases, oil company road maps, Automobile Association maps, county and city maps, business maps and special maps produced for commercial purposes.

There are a few assumptions which we shall dwell on briefly, simply because there are curious aspects involved.

First, we have a common expression: "To orient oneself," a verbal expression in our language meaning to establish knowledge of where you are and in what direction you may be facing. We generally mean: to know where North is, among other things. But the word "orient" means East. We speak of China as being in the far East when it is actually west of us. These expressions are inherited from long ago when people in England and Europe established meanings for all of us, and there was a time when the idea of orienting yourself actually meant to know the direction of the sunrise or the Orient; and most old maps in pre-Columbian time show a circle at the top indicating the rising sun. Now, since the advent of the compass as a common guiding instrument, we have substituted north for east but we still use the old word, "orient," and no one is particularly confused.

The internationally accepted method of pinpointing a location anywhere on earth is based on the assumption that the earth is "round," which it more or less is. We use meridians, north/south lines between the established "true poles" and we start with such a line passing through the metropolitan borough of SE London called Greenwich, England, where King Henry VIII and Queens Mary and Elizabeth were born. There is a little hill, about 180 feet high, in a part of the city called Greenwich Park with a very specific spot chosen in 1675 by Sir Christopher Wren, respected in his day as a mathematician, but now remembered for his design of churches. This is the official point from which longitude is calculated for every place on earth; and the Greenwich Meridian is the north-south baseline for Standard Time, which we refer to as "Greenwich Mean Time."

East-west lines, called parallels of latitude provide us with the other dimension required for location. Since Greenwich is at an odd location in this mode, it has no particular significance. Quite properly we start at the Equator, the locus of the earth's greatest diameter, half way between the North and South Poles.

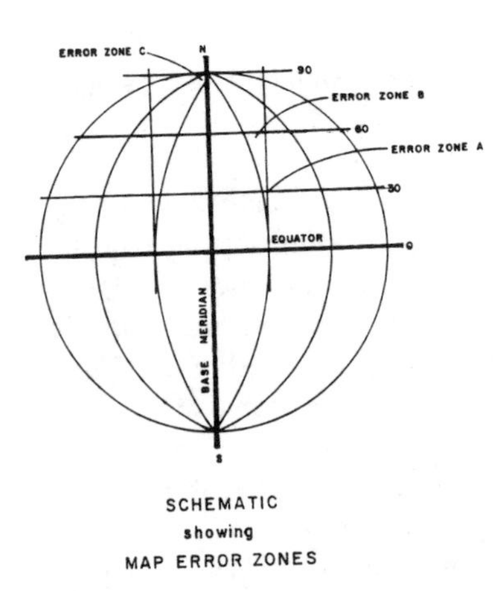

SCHEMATIC
showing
MAP ERROR ZONES

In this schematic, the curved lines represent meridians on the spherical earth. The horizontal lines represent the parallels of latitude. The vertical lines represent meridians as they are ordinarily shown on a map.

There is no problem with the parallels. Near the center of the system, where the equator and the prime meridian meet, there is little meridional difference. But, as we go north, the error increases as shown.

At A, latitude 30, roughly near El Paso, Texas, we can deal with it. At latitude 60, near Skagway, Alaska, however, the error shown at B is substantial, while in the vicinity of the North Pole, C, we have many problems suggested by the drawing showing a meridian not quite making it at all!

The magnetic compass gives a man fits in the far north, and a sighting on the North Star is straight up, making it almost useless in laying out a horizontal line on the ground.

Using angular units (Degrees °, minutes ′, and seconds ″) we measure longitude from the Greenwich meridian east or west 180 degrees to the International Date Line in the Pacific Ocean; and we measure latitude north or south from the Equator toward the poles.

Conventionally we state latitude and then longitude of a point. Thus, a location for Denver, Colorado, would be approximately N40°, W105°. San Francisco, California, would be located approximately at N38°, W132°. These locations can be refined down to hundredth's of a second if necessary.

Legal descriptions of locations are best made using this internationally accepted system; however, within the United States we do recognize a different system created and perfected during the 19th Century for the convenience of the government, railroads, and civil agencies in dealing with people or institutions wanting to acquire land from the U.S. Government or identify specific parcels for various purposes. This is called the Township and Range System of Surveying. It is Cartesian in principle (after Rene Descartes), assuming flat surfaces and squares superimposed on the earth surface, which we acknowledge is neither flat nor square.

Compensation for the differences between assumed flatness on the ground and the reality of spherics are made by an interesting set of adjustments essentially every twenty-four miles. At the equator these adjustments are barely noticeable, while near the poles they are very substantial. In the mid-latitudes (where we live) they are significant enough to argue over, but easy to accommodate.

In a place like New York City, a few square inches may be so valuable that it would take a Donald Trump promotion to finance a small error, while out in the wilderness we might settle a difference of a hundred yards by a flip of a coin! Unless, of course, we found a one foot wide quartz vein carrying 50% gold, and a cubic foot of material could conceivably be worth a million dollars. Then we would want some precision.

Where most placer gold claims are located, as we shall see by the Township and Range System, superimposed on sketches of the land and rivers varying in accuracy depending on when the survey was done and how, the best basic locations are by a system of metes and bounds combined with the rectangular system.

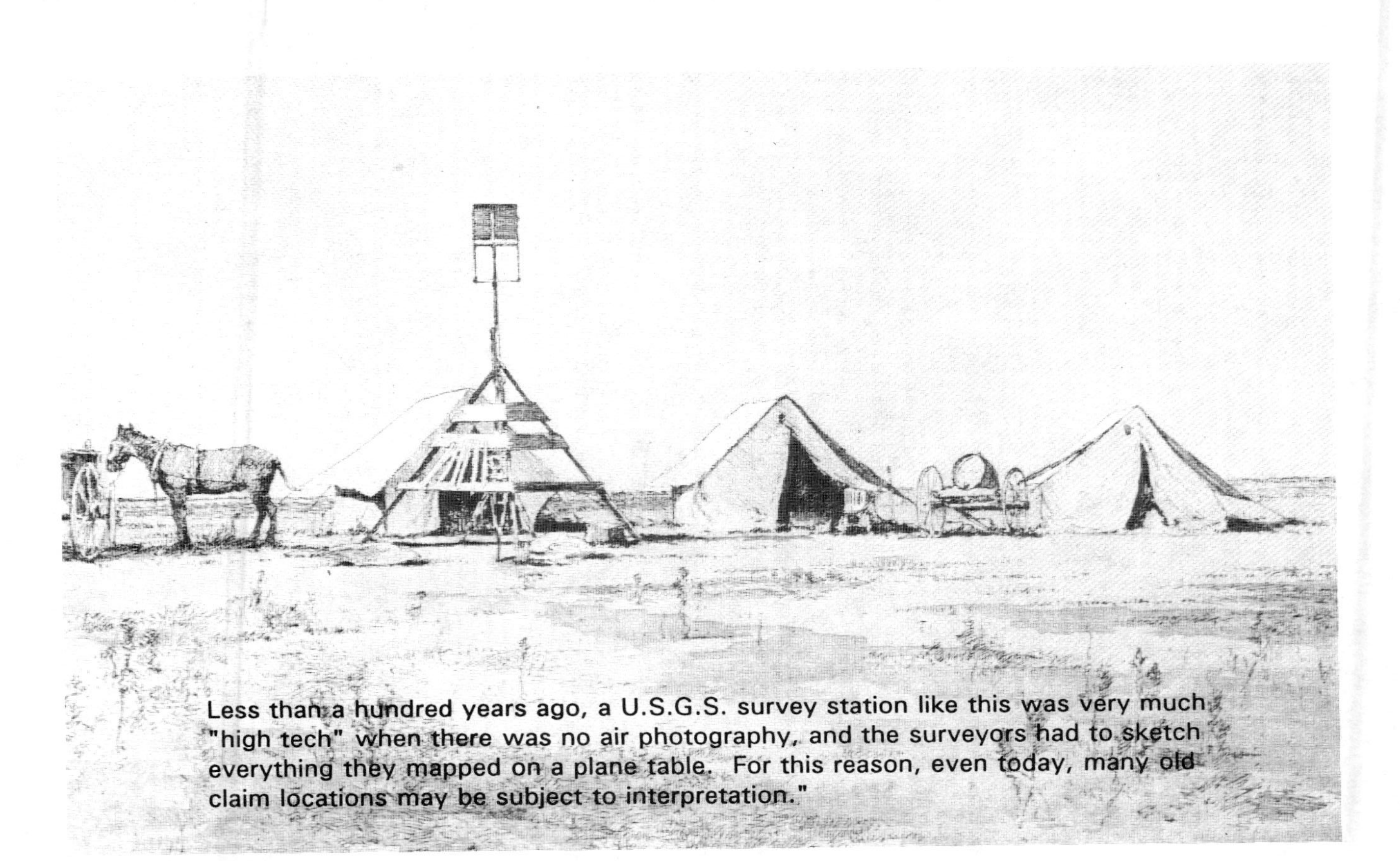

Less than a hundred years ago, a U.S.G.S. survey station like this was very much "high tech" when there was no air photography, and the surveyors had to sketch everything they mapped on a plane table. For this reason, even today, many old claim locations may be subject to interpretation."

There is a space on the official Placer Mining Claim Location Notice form to supplement the ordinary Township-Range description with a metes and bounds description based on markers on the ground; and specific ground details will usually take precedence over generalized locations based on maps alone.

This writer has in his library a copy of the original official Monograph of the United States Geological Survey dated 1893 entitled, A Manual of Topographic Methods which goes extensively into the details of surveying the country up to that time - and would you believe! The State of California is not even mentioned. The nearest is "The Survey of the Fortieth Parallel from 1867 to 1872," directed by a man named Clarence King which, "embraced a zone of country 105 miles in breadth, extending from the meridian of 104 degrees to that of 120 degrees west of Greenwich...." This is essentially a narrow band of country between Denver, Colorado and the Nevada/California line just west of Reno passing through the Salt Lake City area.

The book mentions surveys of the General Land Office covering over a million and a half square miles. This work was largely contracted out to private survey companies and the official book states: "The quality of this work is of varying degrees of excellence."

We learn that the lands and rivers were "sketched" as carefully as possible by people who had to be as much artists as engineers. In many cases the location of rivers was specifically surveyed only on section lines a mile apart and the exact course of the rivers between these points was professionally estimated.

Today's maps, of course, with modern laser beam triangulation, satellite "photography," and high flying air photo surveys is far superior. But the current issue of the U.S.G.S. 1:24,000 Quadrangle entitled DOWNIEVILLE with revisions as late as 1975 shows this writer's mine located with the Downie River on the wrong side of the valley.

A study of the illustration on the next page will prove useful in demonstrating a number of details with relation to claim locations.

First, the four claims which comprise the author's Deep Moon Gold Mine are shown by heavy lines superimposed on this small segment of the U.S.G.S. Downieville Quadrangle.

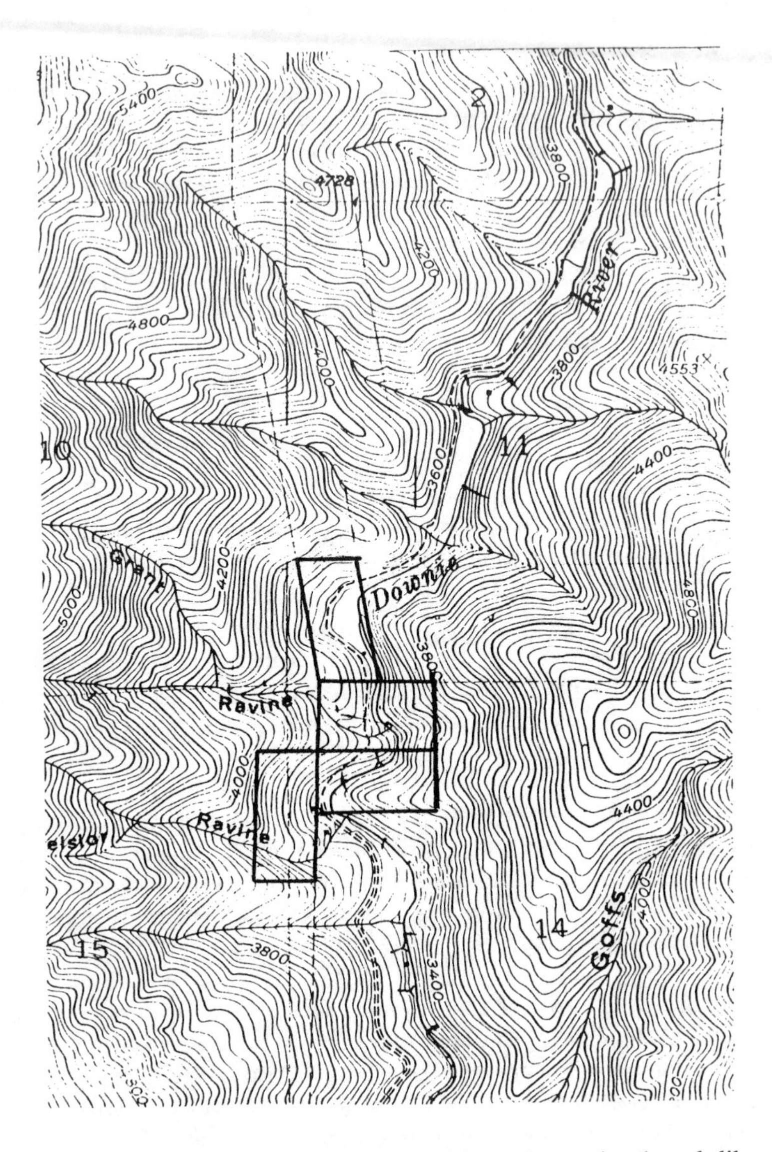

The northernmost of the four claims shown is shaped like a parallelogram not a rectangle, and in the center of that claim the blank white area indicates a flat spot along the Downie River where the mine is located. The river is shown to flow along the toe-head, inside bank, on the east side of the flat area; but, in fact, it flows along the cut bank, outside bank, near the location of the dashed line, which indicates a road that has not been in that location since 1945.

You might get the impression that we are suggesting bad things about the U.S. Geological Survey. Not true. There is little doubt that, given the size of the area and the difficulty of some of the terrain, our available government maps of the western United States are, for the most part, very good. We must simply not take them for granted as precisely correct in all cases.

In the surveying business we speak of errors. (The writer feels qualified to comment on official map errors, having been an instructor in this subject at West Point.) To most people the word error means a mistake, but to a surveyor or an engineer this is not the case. An error may be more properly defined as a departure from absolute correctness in some measurable, but tolerable way.

If you examine the map illustrated on page 34 carefully, you can see numbers starting in the upper right area, then coming down the map and across, then back up. They are 2, 11, 14, 15, and 10. These are section numbers, and they are part of the Township-Range system of indicating locations on American maps. We shall go into the details later, but for the moment let us simply concentrate on errors.

The dashed line running up and down (north-south) half way between the 14 and 15 is, within some limit of error, like a meridian which should continue in a straight line as far as the north pole. But close to the center of the illustration you see the line veer sharply to the west for some distance and then come back to a north-south course. This represents a known error in the system which would probably require a substantial overhaul of data at the highest levels to correct. So we simply don't bother. This represents some kind of adjustment to the facts which we tolerate.

If you think about all this, you can readily understand that this problem in the wilderness has a very low priority in the U.S.G.S. An intelligent professional who could do something about this is going to be much more interested in some dramatic activity like figuring out when the next earthquake may happen along the San Andreas Fault line over near San Francisco, or in studying the explosion inside the Mt. St. Helens volcano fifty miles northeast of Portland, Oregon. And the politicians will authorize millions for government projects like that, conveniently forgetting all about minor defects in the survey system that may bother an old miner back in the woods somewhere.

The authorities will tell you that they are working on these thousands of "errors" throughout the system, and revisions will come along - and they do. But in many instances, we can expect to wait long enough for the river to decide in a storm to change course for the umpty-umpth time and go back to the place where it is already indicated to be on the map.

We must tell you about the farmer living on the east side of the Connecticut River separating New Hampshire from Vermont. His property was on a very fertile ox-bow, which in a big storm was cut off by the river in such a way that he suddenly found himself technically in Vermont! Well, it was only a short time before the tax collector came by to remind him that he would henceforth be paying taxes in Vermont.

When the farmer said that was fine with him, the tax collector was very much surprised. "You know," he said, "the taxes in Vermont are quite a bit higher than in New Hampshire."

"It's worth it," the farmer said, "I never did like those New Hampshire winters!"

So, back to the subject of maps and the Township-Range system which we use in the United States. Bounded by a system of parallels of latitude twenty-four miles apart which are truly parallel, and a system of meridians which have to be corrected because the map lines are designed to be in a rectangular pattern, and the meridians converge making them closer together on the north side of the 24-mile area than they are at the south side, we designate "squares" called Townships, six miles on a side, which contain 36 square miles, each one being identified as a Section.

In a map region, a base point is selected, and the Townships designated to be north or south; east or west of that base point. In northern California, the base point is on Mt. Diablo a few miles southeast of the city of Concord, and on map locations we use the initials MDB&M meaning Mt. Diablo Base Line and Meridian. The location of this writer's gold mine is T20N, R10E, meaning it is 20 townships north (about 120 miles) and 10 townships east (about 60 miles) from the Mt. Diablo base point.

Within each township the sections are numbered, starting in the upper right corner with Sec. 1 and ending in the lower right corner with Sec. 36. Then we have a neat system of subdividing the sections, where you can have "the north half" or "the south half," "the east half" or "the west half."

You can have quarters: "the northeast quarter," "the northwest quarter," "the southeast" and "the southwest quarter." Beyond this you can have halves or quarters of halves or quarters down to very small pieces of land.

Now looking at the illustration showing the four Deep Moon Gold Mine Claims which total eighty acres, let us look at the one we mentioned earlier: the parallelogram with the blank spot more or less in the center. You can see it lies southwest of the number 11. The section number is almost always printed in the center of the section on U.S.G.S. maps. If you imagine a northsouth and eastwest line passing through the number 11, you see this claim is located in the southwest quarter of the section. As a matter of fact, it is in the southwest quarter of the southwest quarter. And finally it may be precisely said to be the west half of the southwest quarter of the southwest quarter of Section 11, T20N R10E of the MDB&M.

The two claims immediately south of the parallelogram, located in Section 14 comprise the northwest quarter of the northwest quarter of that section. We shall leave it to you to figure out the legal description of the fourth claim located in Section 15.

One soon discovers that in reading locations, it is most convenient to read them backward. Thus the last claim, the one in Sec. 15, would be located as "The southeast quarter of the northeast quarter and the northeast quarter of the southeast quarter of the northeast quarter of Sec. 15, T20N, R10E of the MDB&M."

By reading backward we first know where the base point is, and then the Township and Range location, then the Section within the Township, etc. So the whole system turns out to be "slick and handy"! There is a tiny black spot just west of the center of the east side of this last claim. It is an old miner's cabin left over from more than fifty years ago. You might have some fun describing that location using this Township-Range system. It can be done.

These U.S.G.S. maps come in various scales: the most common being 1 inch = 1 mile, 1:62,500. These can show considerable detail and still cover a good sized overall area. There is a more detailed series at a scale of 1:24,000 or about 2-1/2 inches per mile. Then there are other scales all the way to depicting the whole United States on a piece of paper 8-1/2 x 11 inches or smaller like the map on the inside of the front cover covering several hundred miles per inch.

We have one other kind of map to consider because it is so very useful to anyone moving about in the national forest. These maps are published by the U.S. Department of Agriculture. The Tahoe National Forest, where the little town of Downieville, California and the Deep Moon Gold Mine are located, contains about 700,000 acres of public land and 500,000 acres of privately owned land. The map illustrates these ownerships by coloring the public land green. You will immediately note that much of this land is owned in a kind of checker-board pattern, following old laws set forth by Congress. Awarded as incentives to railroad companies or utility companies to encourage them to participate in developing the vast open lands in the last hundred and fifty years or so, or to individuals for cash or rewards for service of one kind or another, or under homestead laws, these lands are technically open to "dealing." One soon learns, however, that the big blocks owned by railroads or electric power companies are essentially bureaucratically controlled and, for all intents and purposes, out of range for the average individual.

N.B. We have said that the U.S.G.S. system of surveying involves corrections every twenty-four miles, and this is generally true. But you will note on the Tahoe National Forest Map that the Standard Parallels (correction lines) are actually thirty miles apart. We have said that the local base point for Northern California is at Mt. Diablo and is called the Mt. Diablo Base Line and Meridian. For the record, there are three Township and Range systems in California: Mt. Diablo in the center/north; San Bernardino in the south, and Humboldt in the northwest.

National Forest maps may be obtained at U.S. Forest Service offices and other locations.

U.S.G.S. maps may be obtained in many sporting goods or hikers equipment stores, among other places. General information may be obtained from the National Cartographic Information Center, Geological Survey, Reston, Virginia, 22092.

Placer mining with dredges on the North Yuba River near Downieville, California. Note the raft at the left has a nice umbrella for shade. Just to the left of the two men standing you can see a girl ready to slip in for a swim.

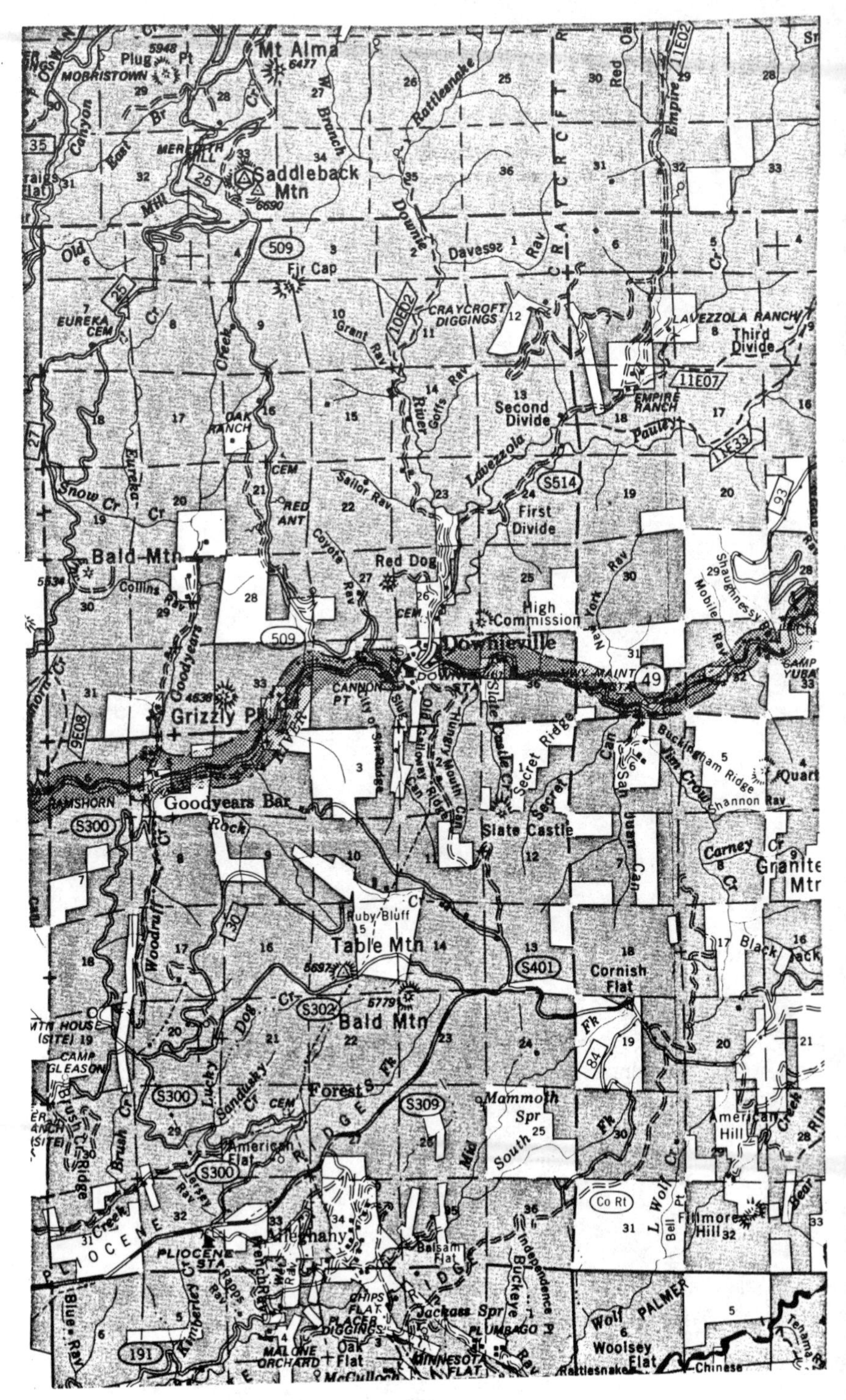

Mt Alma
Plug Pt
MOBRISTOWN
Canyon
East Br
Mill
MEREDITH HILL
Saddleback Mtn
W Branch
Rattlesnake
Downie
Davesse Rav
Empire
Fir Cap
EUREKA CEM
CRAYCROFT DIGGINGS
Grant Rav
LAVEZZOLA RANCH
Third Divide
Goffs Rav
River
Second Divide
EMPIRE RANCH
Lavezzola
Pauley
OAK RANCH
Eureka Cr
Snow Cr
Sailor Rav
First Divide
RED ANT
Bald Mtn
Collins Rav
Coyote Rav
Red Dog
High Commission
New York Rav
Shaughnessy
Mobile Rav
Goodyears
Downieville
Grizzly Pk
CANNON PT
Slate Castle Cr
Secret Ridge
Buckingham Ridge
Jim Crow
Shannon Rav
Goodyears Bar
RAMSHORN
Rock
Slate Castle
Carney Cr
Granite Mtn
Woodruff
Ruby Bluff
Table Mtn
Cornish Flat
Black
Dog Cr
Bald Mtn
Lucky
Sandusky Cr
Forest
Mammoth Spr
South
American Hill
CAMP GLEASON
Brush Cr
American Flat
Mid
L Wolf Cr
Bell Pt
Fillmore Hill
PLIOCENE
Alleghany
PLIOCENE STA
Balsam Flat
Independence
Buckeye
Kimberley Cr
CHIPS FLAT
PLACER DIGGINGS
Jackass Spr
PLUMBAGO
Wolf
PALMER
Woolsey Flat
MALONE ORCHARD
Oak Flat
McCulloch
MINNESOTA FLAT
Rattlesnake
Chinese

The map shown on the opposite page, from the Tahoe National Forest Map, will give you an idea how it compares to a U.S.G.S. Quadrangle.

Comparing this to the map on page 34 you can find Sections 11, 14, 15, and 10 about four miles north of the town of Downieville. Each square is approximately a mile on a side. That is the basic idea, and you will see that the squares near the top of the map are very regular looking, but about an inch down from the top of the map there is a unique dashed line running across the paper, just below the word Saddleback Mtn. This is the Fourth Standard Parallel north of the Mt. Diablo Base Line, discussed earlier on page 36. Major adjustments are made along these standard parallels, and here you can see the sections immediately south of this line are short in the north-south dimension, perhaps only half a mile when they should ideally be a mile long according to the ideal plan.

A glance around this map will reveal sections of very different shape and size with some very large sections near the bottom of the map. Much of this "error" (tolerable departure from the facts on the ground) is probably due to the difficult topography, and the problem of conforming aerial photography to old survey data on the ground.

Going back to Sections 11, 14, 15 and 16 you can transfer the location of the Deep Moon Gold Mine from page 34 to this map and see how it is situated with relation to public and private land. The public land is shown in green on the original Tahoe National Forest map, but it shows up as slightly shaded here. About a mile east of the Deep Moon location you can see a light colored area which is just right of the words Craycroft Diggings. This and another area which is long and narrow, trending more or less north-south about a mile further east, represents a lode claim which has been patented. You can see that they are very irregular as we noted in Chapter 3.

Near the bottom of the map around the little town of Alleghany there are several patented lode claims. Here is the location of the famous 16-1 Mine. And almost exactly half way from Alleghany to Downieville you can see Table Mtn. with Ruby Bluff. This is the location of the famous Ruby Mine, the source of the multi-million dollar exhibit of gold nuggets in the U.S. Mint in San Francisco.

Downieville, California as depicted in a sketch probably made in 1852.

Downieville in the late 20th century with a population of 325.

5. *LOCATING CLAIMS and LEGAL MATTERS...*

Whether you find and locate your own original mineral claim or you buy it from someone else, you need to make certain that the claim is properly identified according to the rules and regulations set out in BLM (Bureau of Land Management) instructions and in compliance with appropriate state regulations.

The first step is to check the claim on the ground:

1. There should be a clear and conspicuous notice in the form of a sign which states the name of the claim, your name and address, the date you discovered/located it, and a brief description of the claim boundaries. In the case of a placer claim, as we have noted, you may simply use the Township and Section system as outlined in Chapter 4.

   A lode claim requires more detailed work on the ground because the rules require that the long axis of the claim be located along the outcrop of the mineral bearing body (usually some kind of vein or other distinctive rock feature).

   Having identified your actual discovery location, you then measure either way, along the strike of the lode (the outcropping of the mineral bearing body) a total of 1500 feet. The discovery point may be, at your discretion, anywhere along the 1500 foot line from one end of the claim to the other, depending on your estimate of the situation.

2. Place 4 x 4 inch posts or the equivalent at the ends of your center line with appropriate markings.

3. Traverse along lines as nearly perpendicular as possible 300 feet in each direction from each stake to locate the claim corners.

4. It is important that you comply with both state and federal law to insure that your claim will be valid.

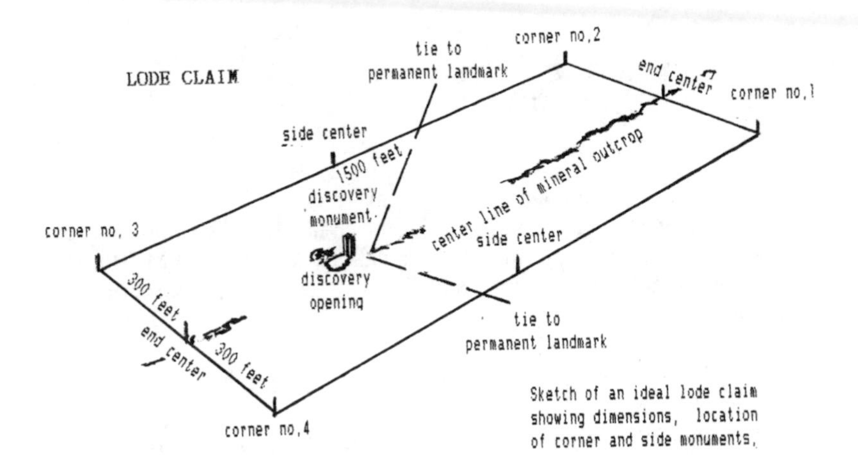

Sketch of an ideal lode claim showing dimensions, location of corner and side monuments.

DISCOVERY MARKERS

Notice as prescribed by law placed in weatherproof container on the marker or protected in the base.

4"+ diameter pole 4' high planted in hole or supported by rocks

large distinctive rock on top of ordinary cairn

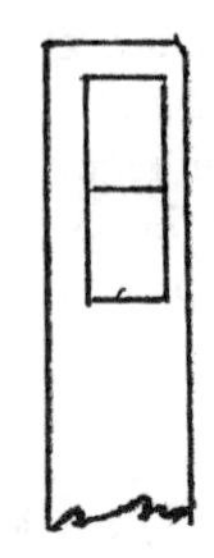

4x4" post with folded metal sheet to protect location notice

CONVENTIONAL CORNER POST

ordinary 4x4" post 4'+ high with appropriate markings

D E T A I L S

These drawings will give you an idea of the general nature of the markers and notices required and the distribution of these on the ground. There may be many variations, but the primary idea is that the location must be clearly identified in a way that avoids confusion.

You may be required to clear, or at least blaze a straight line between your corner posts around the outside of your claim. Here it is quite easy to make a lot of extra work for yourself and a mess of things at the same time. If you are in brush or dense undercover or in steep, up-and-down country, you can soon find yourself far off course, and the upshot of all this will be that there are blazed lines all over the forest.

We shall assume that you have a decent compass and know how to use it, and that you have a 100 foot steel tape. These are basic to the task.

The most important thing for ease and success is to establish that 1500 foot center line and place good posts or cairns at the ends. Having done this, you need to establish clean right angles so you can measure off the 300 feet each way from the ends.

There are two simple ways to set off a right angle in the field, illustrated here and on the next page: 1. Use the basic geometry of a 3-4-5 right triangle by measuring off any multiple as shown. The most common is 6, 8, and 10 feet. Another way is to extend your measured line a short distance (A) beyond your turning point, then swing arcs (B) from equal distances (A) on either side of the corner to intersect on the perpendicular as shown.

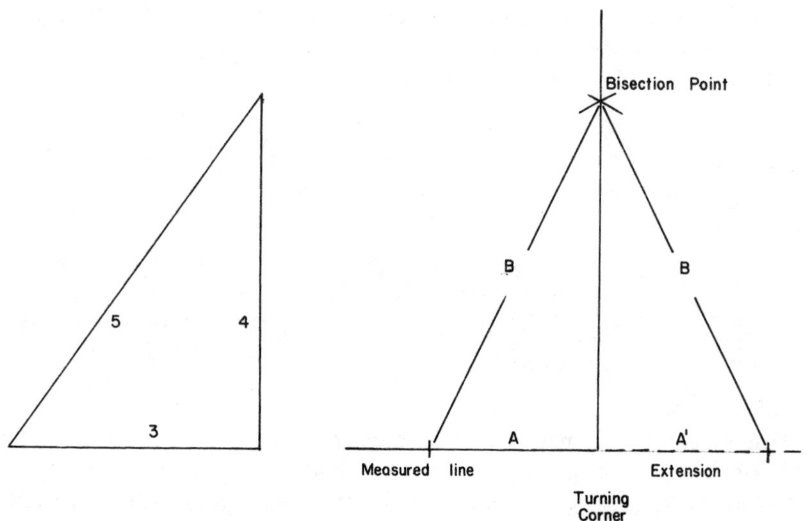

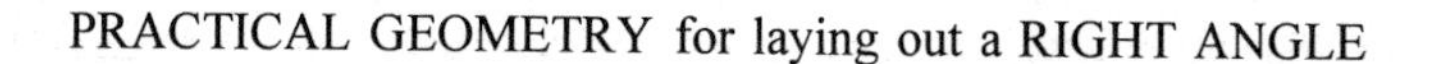
PRACTICAL GEOMETRY for laying out a RIGHT ANGLE

Two miner friends of the author demonstrate use of a steel tape in laying out a right angle. Even a rock comes in handy.

Two practical rules apply:

1. After establishing your center line, do the shortest lines first.
2. Do not set permanent markers until you have done the short lines with temporary markers or flags, so you can make adjustments if you find you have wandered off course a bit.

It may take an extra hour or two, but it will save a lot of time in the long run if you establish the center of your center line, that is the 750 foot point. You may need centers of the long side lines anyway, so this could be good for more reasons than one.

Measure a perpendicular 300 feet away from the center point on the center line each way. Now you have established a good check point on each of the two long outside lines, so instead of having to work your way 1500 feet before you know if you are on line or not, you only have to go 750 feet.

In attempting to follow a straight line over rough ground or in places where you can not see very far at ground level, you should always try to make the longest sighting that you can. If at all possible, pick a point on the horizon to keep in your sights, and better, pick two that you can keep lined up as you move along your course; then look back to confirm that your line is truly straight.

Having located your claim on the ground, the next step is to do the paper work and file your location with the authorities:

1. The proper form is entitled PLACER MINING CLAIM LOCATION NOTICE, Form Number CSO 3800-1 or LODE MINING CLAIM LOCATION NOTICE, Form Number 3800-2.

2. Record your claim in the county of the state where it is located.

3. File this recording with the nearest BLM office.

4. Comply with the annual requirements to keep your claim "current."

Some details regarding assessment work and taxes have changed in recent years or may be subject to change in coming years. It is your responsibility to keep up with these as they become effective.

The law covering mine claims can be very complicated and confusing. For example, let us consider two lode claims side by side, each one properly following a quartz vein on the surface. The line of the outcrop is called the strike of the lode or vein. The direction it takes down into the ground is called the dip of the vein. Since the original source of this quartz and the valuable mineral encased in it is thought by geologists to have been gaseous or liquid in nature, and probably originating in a single source, it is quite possible that the two separate veins at the surface would dip in such a way that they would meet underground. The meeting point is very important since it is often the location of the richest part of the deposit.

You can see that as you follow these veins down into the earth, it is not only possible, but quite likely that one or both of them will drift away from the vertical to the point of being located outside the claim boundaries at the surface.

In a dispute over ownership between the two claims, the area of vein intersection will belong to the Senior Claim even though it is clearly located under the Junior Claim.

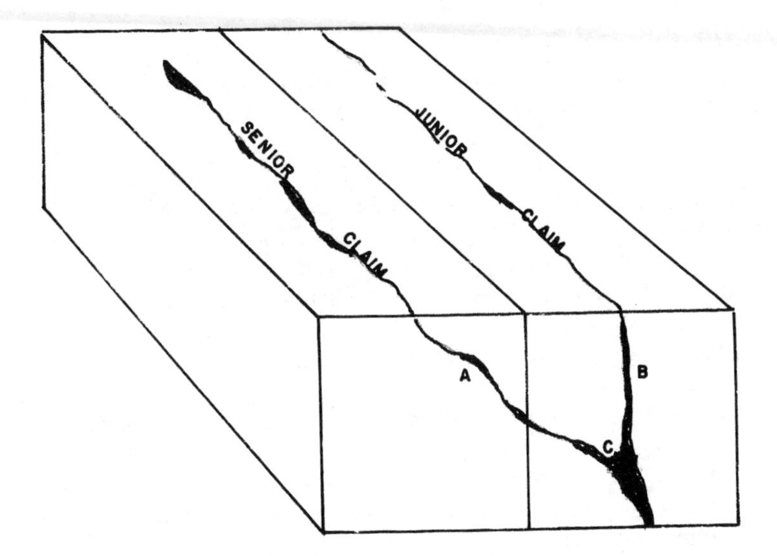

EXTRALATERAL RIGHTS
DIAGRAM

Question: What is a Senior and what is a Junior Claim?

Basically the Senior Claim is the one located first. But the date of recording is also taken into consideration in some cases, and the condition of the claim may also be brought into the dispute. When a great deal of value is at stake, as it can be in the mining business, lawyers will go to extremes to establish this state of "Seniority," and a minor departure from the law such as failing to meet a publication date could be critical.

To illustrate the degree of complication further, if it is found that the Senior Claim has lapsed for failure to do assessment work, for example, it can be proved to be invalid. The Junior Claim can not, however, simply lay claim to the area in conflict without relocating the conflict area during the period when this is permitted. A third party could relocate the lapsed Senior Claim and take the prize even though the third party relocation is subsequent to the Junior Claim.

But, enough of this "scare talk"! The likelihood of your being involved in such complicated litigation is very slim. Nevertheless it pays to keep things straight and in compliance with the law.

We are focused on enjoying yourself and having a place to camp or hide away with or without a few friends at your pleasure; but you must keep in mind that mine claims are not and never were set up for that purpose.

A mine claim is a mine claim first, last, and always. If you happen to be able to swim in the beautiful deep spot in the river, or pitch a tent, or pour a concrete platform for a camper, or open a spring for water, or clear a site for a cabin, all these activities are subordinate to the actual reason for the claim in the first place. If you are ever challenged by a U.S. Forest Service person or a State Fish & Game person or any other person in authority, you had better be able to demonstrate that you are fundamentally involved in a legitimate mining enterprise. It does not have to be a big effort, but it does have to be a demonstrable effort.

Disputes between miners are infrequent, but each year that passes finds those of us who enjoy life in the wilderness more and more subject to the whims and animosities of a certain kind of individual who resents our being in the forest at all. These people can be a real problem at times. In the name of preserving an area or something they don't want changed, they can organize in formidable ways to influence a legislator or a public official. They can also be extremely intimidating in many ways, and we have to be very careful how we behave and what we do, no matter how frustrating or wrong-headed they may be.

It seems as though country people and people who like to be in the wilderness are generally not disposed to "organize." They are content to leave things alone, for the most part. For this reason they are vulnerable to those who would organize to impose their wills on us by urging legislation to limit our activity in any number of ways. When we do realize that they are about to use force on us in one form or another, and we do react to them, we are still not well enough organized to be especially effective, because we do not think in legalistic terms, thinking of ourselves more as simply following the precepts of common sense.

At this point, without laboring these "political views" too much, it is probably sufficient to point out that most Americans today feel frustrated in many, many ways where the acts of legislators seem to reflect a complete absence of common sense in dealing with the various problems in society. All they seem to be able to do is spend money, and make laws that limit what we can do.

As an example, seat belts really do not make us better drivers. If they did, laws requiring them might be a good thing. We have those laws only because someone thinks it is good for us, so they force a legislator through lobbying efforts to vote in a law that imposes their will on us - their idea of what is best for us, whether we want it or not. Fortunately, in an event like this, the people required to enforce such a law, at all levels, tend to be rather lenient about it, so very few of us are brought to task for not having a seat belt on, except when those "do-gooders" start monitoring the public officials and forcing them to force us to submit to their will - common sense or no common sense.

In picking the issue of seat belts it is not this writer's intention to argue that particular point one way or the other. It is only one of hundreds of such cases where a loud voiced minority is able to force its will on a simple, but basically complacent majority.

In the northwestern United States, in recent years, many of the people who would impose their wills on country people have become dangerously militant. In opposing logging, they have driven good sized spikes into trees in such a way that when a legitimate lumberman, operating well within the law, attempts to cut down and limb a large tree he can hit one of these steel spikes with his chain saw and seriously hurt himself when the saw jumps out of his control or if the tree arrives at the saw mill, the big circular saw can hit the spike causing even greater damage and injury to workmen.

Another favorite way for militant environmentalists to make trouble is to sneak onto a mining or lumbering property and pour foreign material such as sand or sugar into the fuel tank of a big piece of diesel driven equipment. If that isn't enough they can charge by a working operation at good speed, because their efforts are almost always cowardly in nature where they dare not confront a working country person head on; and stop just long enough to shoot a round or two from a large caliber rifle into a diesel fuel tank, spilling the stuff out on to the ground. They have many tricks up their sleeves to intimidate country people.

The basic Claim Notice may be a very simple affair.

We must remember that they are not only active in the field, but they are very active in the halls of the legislatures and in the courts. In all these ways they can be extremely devious, so you have to be up early to keep them from tricking you with a new law.

As an example, when you file a mine claim, it is on federal land. That is the only kind of land where you can do this. Now, there is a large network of existing roads in varying condition already in the National Forest or other public domain. As a mine claim owner you have an absolute right to use these roads, but these people who want you out can damage a road to your operation again and again until the U.S. Forest people stop fixing it or take their time doing it, and you can not afford to do it all by yourself. And, if this won't stop you, these people will lobby with great power to change the basic law that has been in effect and worked very well for over a hundred years.

There are two primary statutory sources of your right: The Mining Law of 1872, and Title V in the Federal Land Policy and Management Act of 1976. Under the implied right in the 1872 law you don't have to apply for a right of way or pay rent for a road. The act is subject, however, to the surface use regulations administered by the BLM, which are designed to minimize the impact of mining activities on the surface resources. Quite properly you have to give notice, and then guarantee appropriate reclamation after you are finished. A review of your plan may involve some kind of impact study including environmental factors, archeological considerations, wildlife surveys, hydrological surveys, or sociological impact studies, depending on the size and scope of your operation. The size and scale of effort we contemplate in this book hardly involves such things as this. But sometimes a large operation can lock heads with a group of adamant environmentalists and a law can emerge which may affect all of us in some blanket, adverse way that would otherwise have never come up.

The Sierra Club was at one time a wonderful organization of people who actually knew each other and liked to go out hiking or even bike riding. But over the years people began to infiltrate the club with heavy handed political objectives and all us old hikers decided to give up our membership. This writer has not been a member since the early 1970's because the Sierra Club has moved in such a force-handed way into the political arena with activities that defy any semblance to common sense.

Today the Sierra Club is openly out to get us all out of the National Forest by 1993 with the slogan . . . "Mine free by '93."

The Mining Law System, in effect for well over a hundred years, as American as apple pie, is an organization of law covering the land as a primary source of wealth. The strength of our country is in the concept of private ownership of land.

After years of abuse of land and people's individual rights, the Soviet Union, Eastern Europe, and China are involved in most painful efforts to rediscover these rights which we take for granted in the United States of America.

With our private enterprise system we enjoy a standard of living where we can afford to pay for sensible environmental protection in all ways and at all levels. But we can not allow idiotic efforts to undermine the basic ideas of Americans to save some bird or snail or engage in some other activity totally devoid of good old common sense.

We are just now learning that governments that have not allowed their people to function as individuals, have polluted their part of the world beyond belief in the name of government planning and control.

These are serious matters and, if we are to continue to enjoy an opportunity as individuals and as families to go out into the national lands and "poke around" looking for gold or the mineral of your choice, we must be vigilant because there are those out there who would just as soon run your car off the road as look at you.

One last point on this issue and we will go on to more pleasant aspects of this subject.

Specifically, there is an organization called Earth First which has a paperback written by one of their leaders entitled "Ecodefense: A Field Guide to Monkeywrenching." In it you can find these words:

> *"If loggers know that a timber sale is spiked they won't bid on the timber. If a Forest Supervisor knows that a road will be continually destroyed, he won't try to build it. If seismographers know that they will be constantly harassed in an area, they'll go elsewhere. If ORVers know that they'll get flat tires miles from nowhere, they won't drive in such areas."*

This is a clear case of preaching terrorism in out and out defiance of the law and the American way of life. Earth First activists are well known for attacking in stealth and at night, destroying property and machinery, endangering human life and then fleeing from the law.

The record shows that Earth First opposes nearly all use of the public lands and openly advocates and practices sabotage, violence, and terrorism. They have directed their violence against timber people, cattlemen, woolgrowers, miners, oil and gas prospectors, ski resort owners and off road vehicle people. They have so much power and are so much feared that the U.S. Forest Service seems to shrink from dealing with them.

Dealing with a recent problem in the George Washington National Forest in Virginia, we are told that George W. Kelley, supervisor of that forest not only agreed to negotiate with Earth First!, but placed one of their representatives on an official U.S. Forest Service panel to participate in developing a forest plan. When he was reminded of the terrorist character of that organization by William Perry Pendley, President and Chief Legal Officer of The Mountain States Legal Foundation, he paid no attention. Mr. Pendley then filed an appeal to F. Dale Robertson, Chief of the U.S. Forest

Service, who simply said words to the effect: "We haven't had any trouble with them." Mr. Pendley responded by arguing that "the U.S. State Department hasn't had any trouble with the Islamic Jihad in New York City but that doesn't mean that it has any right to negotiate with that terrorist group there."

Finally, when Mr. Pendley, with maximum persistence, asked the Chief to send him a letter confirming the official position of the U.S. Forest Service saying, "I want you to explain to the American people why the U.S. Forest Service is negotiating with a terrorist group," he received a letter six weeks later saying that the U.S. Forest Service had disbanded the entire panel.

Mr. Pendley concludes his report on this matter saying: "The decision of the Forest Service is also a lesson to all of us that we can fight back and that we can win. It is a lesson we all need to remember."

Such talk as this may discourage you from trying to find that mining claim where you can at least enjoy some time in the open country, and truly have a chance of discovering a really valuable deposit of precious metal. If you want to let terrorists and well meaning do-gooders with no respect for common sense take the open country in America away from you, well so be it. But we shall continue on in the hope that you are not going to be put down that easily and can at least vote for the right people while this problem is in the bud.

6. *PRINCIPLES of DEALING and ESTABLISHING VALUES*

How much is a mine claim worth?

There are several conventional ways of establishing the value of any piece of real estate; and a mine claim is a special form of real estate.

The first thing to do is define how much right, title, and interest an individual will have in the property.

In the case of a Patented Claim this is as good as it comes. It is real property, held in fee simple, direct from the United States of America. Almost anywhere in this country 20 acres of that kind of land will be worth a lot of money. We will go into specific evaluation a bit later.

In the case of an Unpatented Claim, the ordinary run-of-the-mill variety, we have two kinds:

1. A claim which has a proven discovery can be legally and formally "located." This gives the claim some value above that of most unpatented claims. But, there are very few of these; and the extra value, while arguable, is somewhat nebulous.

2. One of the almost 1,500,000 recorded mine claims in the western United States that do not have a discovery as defined in the mining laws. This is the kind of claim we are talking about for all intents and purposes.

The fundamental question is: What can you do on a property like this? It is far from regular real estate but, nevertheless, you do have well established Prediscovery Rights or Pedis Possessio.

Over the years, in many court decisions it has been acknowledged that the requirement for discovery to precede location is impractical. The reason for this is that most valuable minerals are hidden below the surface in one way or another and require considerable time and expenditure of money to prove their existence; and no one in his right mind would be willing to spend a great deal of money unless he had some protection from the general public.

If a claim were not staked or publicly claimed in the records prior to conducting an exploration program, rival claimants hearing of the exploration activity might locate claims over the area.

The doctrine of pedis possessio which literally means "a foothold" was established to provide such protection. This possessory right is limited to protection against adverse locators or the general public which is, of course, all you really need.

As a claim holder you have a right of access, but this is not in any way exclusive. If, however, there is a question of liability, a fence may be erected and multiple locks employed allowing entry only to the government and the claimant.

You may put in power lines or water conveyance devices without a right-of-way permit.

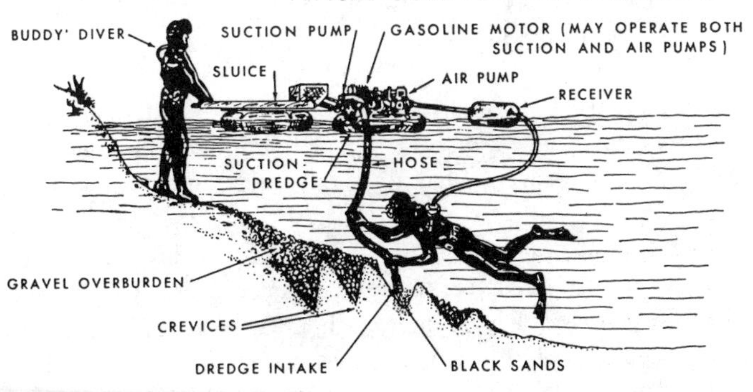

In the typical gold diving operation, the gasoline-engine powered air compressor and suction pump can be located on shore or mounted on flooats so they can be towed by the diver as he works under water. (From "Diving for Gold" by William B. Clark, California Geology, *April, 1972, California Division of Mines and Geology.)*

There are certain "grandfather conditions" which can become a bit sticky, such as exclusive automatic permissions to claims located prior to July 23, 1955.

Basically, as long as you are only prospecting, you probably will be limited to having a tent on a temporary camp site. You can, in most instances, get away with some kind of secure storage building to keep things on site during the off season or periods of inactivity but that is all.

When you can say you are mining (which may be any reasonable scale of activity to fit the local conditions) you may then erect buildings or cabins for residential purposes as long as this is directly related to mining activity and it can be demonstrated that you need to live on site when the mine is operating. The key idea is that buildings may be legitimately placed and maintained on an unpatented claim by the original locator and all subsequent owners providing that such buildings are essential to the mining operation.

The government frowns on just having a cabin on a claim with obviously no mining going on but, if you have a reasonable thing going, it is obvious that you would not want to be away overnight where someone could rip off either your equipment or your gold.

There is no way you can acquire any rights against the government by prescription or adverse possession just because you have occupied a claim continuously for a long time or anything like that. The only way to get perfect possession is to patent the claim.

It is generally true today that as long as you are simply prospecting you do not need more than the ordinary permits such as permits for campfires or for dredging, which are usually quite easy to get and do not cost a great deal, if anything. It is important to get such permits each year and comply with their regulations.

When you elect to start mining, this may be a very small operation but you will need to apply for a mining permit with the U.S. Forest Service, and you may immediately find yourself dealing with a long list of agencies including the Fish & Game people, Water Quality Control, even OSHA if you have any helpers or employees, certain county officials such as Health Department people, solid waste disposal or environmental protection authorities, etc., but this is not greatly different from any rules and regulations now imposed on us all in society.

Curiously, the statutes contain some fascinating details:

If you abandon a claim, whatever you leave behind automatically becomes the property of the government and may be taken over by a new locator, except in the case of a "treasure trove." A treasure trove is defined as "any gold or silver in coin, plate or bullion found concealed in the earth, or in a house or other private place but not lying on the ground, the owner of the discovered treasure being unknown, or the treasure having been hidden so long as to indicate that the owner is dead." Confusing? Not to a sharp lawyer!

When a claim is abandoned, the real property reverts to the government but the personal property, unless specifically claimed by the government through a process called "asserting dominion," may become the property of the finder. This would include vehicles, tools, or machinery, etc. So, if you find something that is not real estate (attached firmly to the ground) on an old mine claim, you can probably take it home and call it yours.

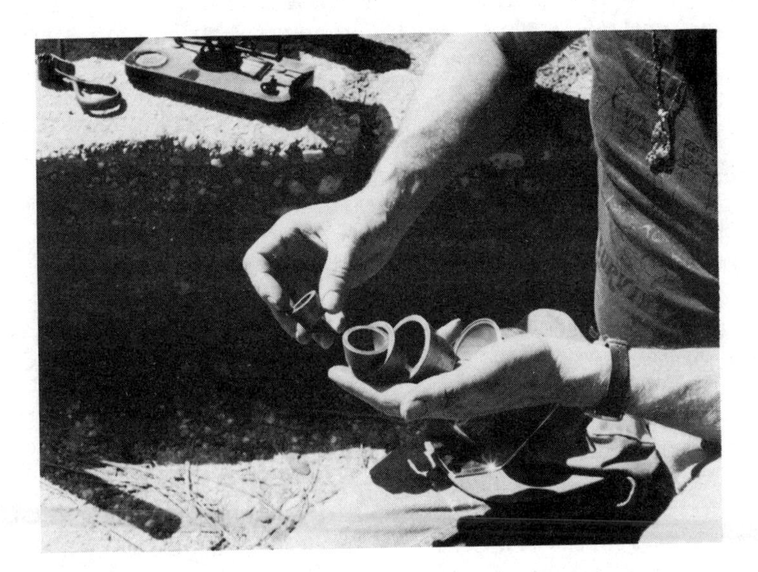

Dave Jeffries of Allegheney, California displays some of the museum quality small objects he has found on old mine sites or at local yard sales.

Please remember that, even if you think you have the right, if you do not, and you take something off a properly located mine claim, you may be liable for unlawful use, removal or damage to abandoned property and find yourself involved with the Criminal Code and a fine of up to $10,000 or imprisonment of up to 10 years. But, not to worry - legal interpretation indicates that "not only must government ownership and a trespassory taking be shown, but it must also be shown that the defendant acted with an intent to appropriate property he knew to belong to another."

If, on the basis of this information so far, you can determine what you can and can not do on an unpatented mining claim, you are in some kind of position to render an evaluation in comparison to ordinary real estate. It adds up simply to the idea that you can treat the property just like regular real estate as long as your activity is clearly tied to actual mining, no matter how small the enterprise, and you can hold the claim for years and years and pass it on to your heirs, successors, and assigns.

There are even some implied advantages over real property rights. For example, in much isolated country, if you owned a parcel of land and you wanted to construct a residential building, you might find the county authorities very difficult to deal with because they know that they have some obligation to send the Sheriff out or the Fire Department if you have trouble, and they can not afford that on the basis of the relatively small tax that may be imposed on you; and they do not like that. All rural counties in California and, probably in all the western states, are constantly short of funds because they have such a small tax base (so much of their land is federal and not subject to local taxes); and they have an inordinate number of people on unemployment compensation or welfare. If you have a mine claim and you want to build a residential building, the county has no automatic obligation to serve you in the same way that they do when you own the ground. They can not object to your cabin, but you may be sure they will tax it and probably hit you for support of the local dump (solid waste disposal area) whether you ever see it or not. But that's the way things are and, like all governments, they are going to collect all the taxes they possibly can.

Now, getting on to evaluating that mine claim, let us consider some of the details as the professionals see them. You may or may not bring them all into play, but it is worthwhile to know about them.

Here are the accepted Principles of Valuation:

Change What changes have taken place or may take place which will affect the value?

Substitution What would a reasonable substitute cost?

Supply and Demand Price per real value varies directly, but not necessarily proportionately, with demand; and inversely, but not necessarily proportionately with supply.

Conformity Value depends on how much the property is like others in the vicinity.

Highest and Best Use What is the best use that can be made of the property in question?

Progression The worth of a lesser valued object tends to be enhanced by association with similar objects of greater value.

Regression The worth of a greater valued object tends to be reduced by association with similar objects of lesser value.

Contribution Maximum values are achieved when improvements produce the highest return commensurate with their cost.

Anticipation Value is created by anticipated future benefit.

Competition Value creates competition which creates supply which brings about less cost.

Balance Over improvement or under improvement can influence the value to cost ratio in terms of investment.

There are basically three ways to arrive at an estimate of value:

1. Market Comparison Check the price of actual sales of similar property, not the asking prices.

2. Cost Determine what the seller paid for the property by such efforts as checking recordings of sales or taxes, etc.

3. Income Estimate the potential income of the property and capitalize the net income into value. This is what is done with actual mines, operating or non-operating.

You will probably find yourself using a combination of all these ideas in figuring out the value of the mine claim in which you may be interested. Therefore, in the next few pages we display advertisements taken from a recent single copy of the California Mining Journal to give you some idea of indicated values and availability of mining claims all over the west.

John Sieler demonstrates the bulletin board in his gold store in Downieville, California In winter, he is in Quartzite, Arizona.

COARSE GOLD PLACER CLAIM. N.F. Feather River at Belden, CA. Mineral survey #5211 and 5627, 26.78 acres from Belden Bridge, 2400 feet downstream. RV campsites. $8,000 OBO, & others in Idaho. JIM, (503)230-1014. eAApr1

HAVE PLACER MINE with excellent potential. Will entertain joint venture, lease or possible purchase. Must have proven credentials. J-BASH CORP., (619)588-8707. eAApr1

MINING CLAIMS (over 200 acres) in NW California. Over 30 acres of bench (was patented). Year-round dredging. $14,000 obo. Divorce forces sale. P.O. Box 51641, Phoenix, AZ 85076; (602)497-2752. fSApr1

PATENT YOUR PLACER CLAIM! Contact STEVE HICKS, former BLM mineral examiner specializing in placers. (406) 222-1707. eMar2

NORTHERN CALIFORNIA GOLD: 8 Klamath River claims for sale. Proven area. Fine gold throughout. Partnership breaking up. $8,000. JOHN DORNER, 10403 SE Reedway, Portland, OR 97266. aApr3

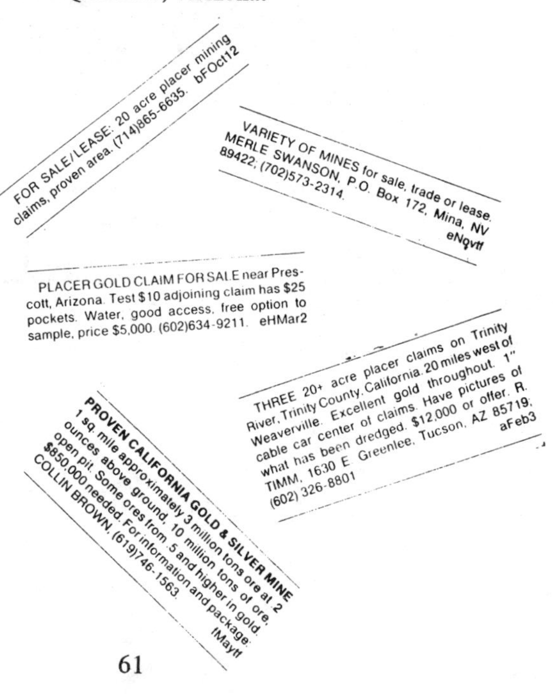

FOR SALE/LEASE: 20 acre placer mining claims, proven area. (714)865-6635. bFOct12

VARIETY OF MINES for sale, trade or lease. MERLE SWANSON, P.O. Box 172, Mina, NV 89422; (702)573-2314. eNovtf

PLACER GOLD CLAIM FOR SALE near Prescott, Arizona. Test $10 adjoining claim has $25 pockets. Water, good access, free option to sample, price $5,000. (602)634-9211. eHMar2

THREE 20+ acre placer claims on Trinity River, Trinity County, California. 20 miles west of Weaverville. Excellent gold throughout. 1" cable car center of claims. Have pictures of what has been dredged. $12,000 or offer. R. TIMM, 1630 E. Greenlee, Tucson, AZ 85719; (602) 326-8801 aFeb3

PROVEN CALIFORNIA GOLD & SILVER MINE 1 sq. mile approximately 3 million tons ore at 2 ounces above ground, 10 million tons of ore, open pit. Some ores from .5 and higher in gold. $850,000 needed. For information and package: COLLIN BROWN, (619)746-1563. fMaytf

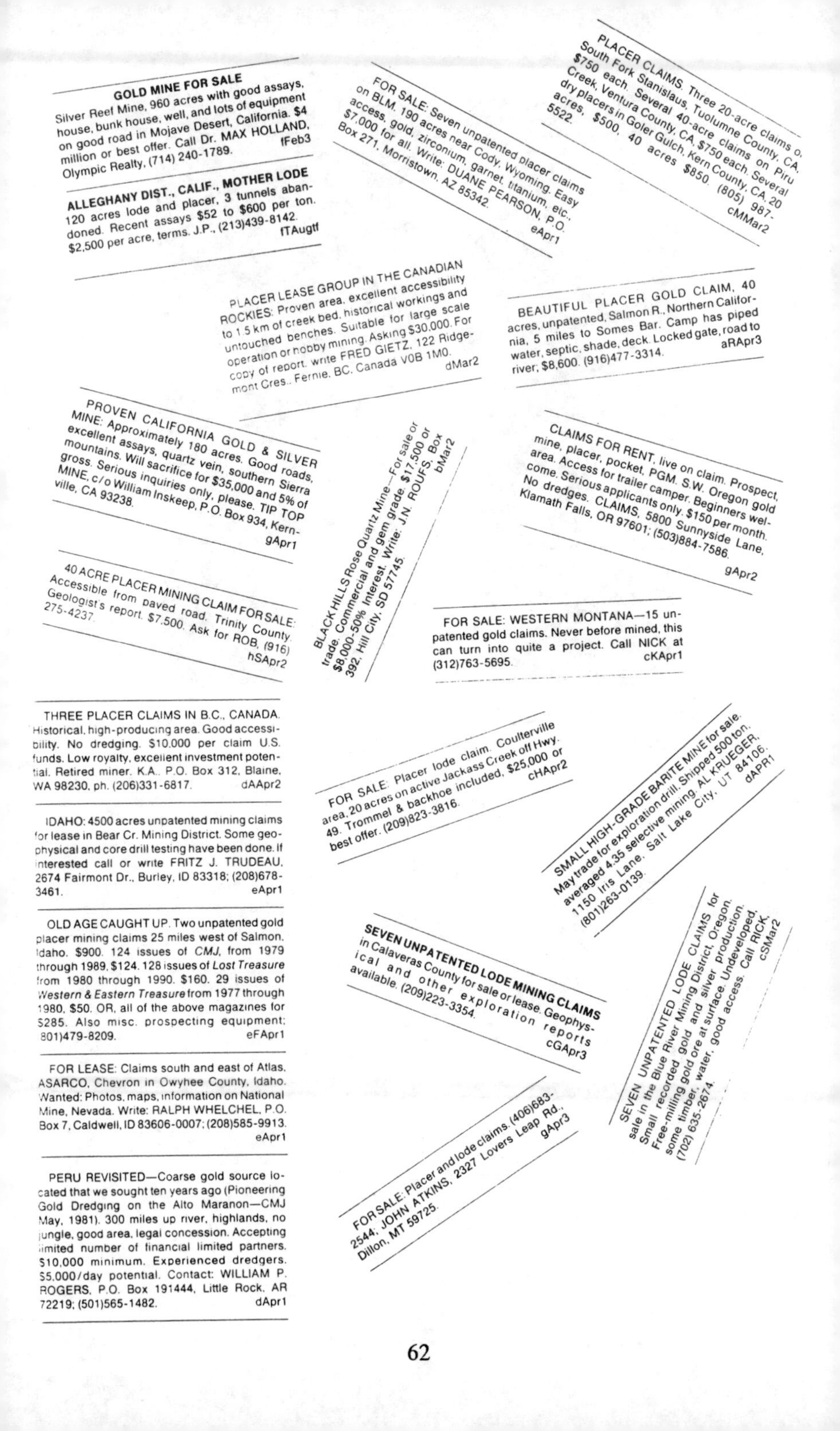

GOLD MINE FOR SALE
Silver Reef Mine, 960 acres with good assays, house, bunk house, well, and lots of equipment on good road in Mojave Desert, California. $4 million or best offer. Call Dr. MAX HOLLAND, Olympic Realty, (714) 240-1789. fFeb3

ALLEGHANY DIST., CALIF., MOTHER LODE 120 acres lode and placer, 3 tunnels abandoned. Recent assays $52 to $600 per ton. $2,500 per acre, terms. J.P., (213)439-8142. fTAugtf

FOR SALE: Seven unpatented placer claims on BLM. 190 acres near Cody, Wyoming. Easy access, gold, zirconium, garnet, titanium, etc. $7,000 for all. Write: DUANE PEARSON, P.O. Box 271, Morristown, AZ 85342. eApr1

PLACER CLAIMS. Three 20-acre claims on South Fork Stanislaus, Tuolumne County, CA. $750 each. Several 40-acre claims on Piru Creek, Ventura County, CA, $750 each. Several dry placers in Goler Gulch, Kern County, CA, 20 acres, $500. 40 acres $850. (805) 987-5522. cMMar2

PLACER LEASE GROUP IN THE CANADIAN ROCKIES: Proven area, excellent accessibility to 1.5 km of creek bed, historical workings and untouched benches. Suitable for large scale operation or hobby mining. Asking $30,000. For copy of report, write FRED GIETZ, 122 Ridgemont Cres., Fernie, BC, Canada V0B 1M0. dMar2

BEAUTIFUL PLACER GOLD CLAIM, 40 acres, unpatented, Salmon R., Northern California, 5 miles to Somes Bar. Camp has piped water, septic, shade, deck. Locked gate, road to river; $8,600. (916)477-3314. aRApr3

PROVEN CALIFORNIA GOLD & SILVER MINE: Approximately 180 acres. Good roads, excellent assays, quartz vein, southern Sierra mountains. Will sacrifice for $35,000 and 5% of gross. Serious inquiries only, please. TIP TOP MINE, c/o William Inskeep, P.O. Box 934, Kernville, CA 93238. gApr1

BLACK HILLS Rose Quartz Mine—For sale or trade. Commercial and gem grade. $17,500 or $8,000-50% Interest. Write: J.N. ROUFS, Box 392, Hill City, SD 57745. bMar2

CLAIMS FOR RENT, live on claim. Prospect, mine, placer, pocket, PGM. S.W. Oregon gold area. Access for trailer camper. Beginners welcome. Serious applicants only. $150 per month. No dredges. CLAIMS, 5800 Sunnyside Lane, Klamath Falls, OR 97601; (503)884-7586. gApr2

40 ACRE PLACER MINING CLAIM FOR SALE: Accessible from paved road. Trinity County. Geologist's report. $7,500. Ask for ROB, (916) 275-4237. hSApr2

FOR SALE: WESTERN MONTANA—15 unpatented gold claims. Never before mined, this can turn into quite a project. Call NICK at (312)763-5695. cKApr1

THREE PLACER CLAIMS IN B.C., CANADA. Historical, high-producing area. Good accessibility. No dredging. $10,000 per claim U.S. funds. Low royalty, excellent investment potential. Retired miner. K.A., P.O. Box 312, Blaine, WA 98230, ph. (206)331-6817. dAApr2

FOR SALE: Placer lode claim. Coulterville area, 20 acres on active Jackass Creek off Hwy. 49. Trommel & backhoe included, $25,000 or best offer. (209)823-3816. cHApr2

SMALL HIGH-GRADE BARITE MINE for sale. May trade for exploration drill. Shipped 500 ton, averaged 4.35 selective mining. AL KRUEGER, 1150 Iris Lane, Salt Lake City, UT 84106. (801)263-0139. dAPR1

IDAHO: 4500 acres unpatented mining claims for lease in Bear Cr. Mining District. Some geophysical and core drill testing have been done. If interested call or write FRITZ J. TRUDEAU, 2674 Fairmont Dr., Burley, ID 83318; (208)678-3461. eApr1

OLD AGE CAUGHT UP. Two unpatented gold placer mining claims 25 miles west of Salmon, Idaho. $900. 124 issues of *CMJ*, from 1979 through 1989, $124. 128 issues of *Lost Treasure* from 1980 through 1990. $160. 29 issues of *Western & Eastern Treasure* from 1977 through 1980, $50. OR, all of the above magazines for $285. Also misc. prospecting equipment. (801)479-8209. eFApr1

SEVEN UNPATENTED LODE MINING CLAIMS in Calaveras County for sale or lease. Geophysical and other exploration reports available. (209)223-3354. cGApr3

SEVEN UNPATENTED LODE CLAIMS for sale in the Blue River Mining District, Oregon. Small recorded gold and silver production. Free-milling gold ore at surface. Undeveloped, some timber, water, good access. Call RICK, (702) 635-2674. cSMar2

FOR LEASE: Claims south and east of Atlas, ASARCO, Chevron in Owyhee County, Idaho. Wanted: Photos, maps, information on National Mine, Nevada. Write: RALPH WHELCHEL, P.O. Box 7, Caldwell, ID 83606-0007; (208)585-9913. eApr1

FOR SALE: Placer and lode claims. (406)683-2544. JOHN ATKINS, 2327 Lovers Leap Rd., Dillon, MT 59725. gApr3

PERU REVISITED—Coarse gold source located that we sought ten years ago (Pioneering Gold Dredging on the Alto Maranon—CMJ May, 1981). 300 miles up river, highlands, no jungle, good area, legal concession. Accepting limited number of financial limited partners. $10,000 minimum. Experienced dredgers. $5,000/day potential. Contact: WILLIAM P. ROGERS, P.O. Box 191444, Little Rock, AR 72219; (501)565-1482. dApr1

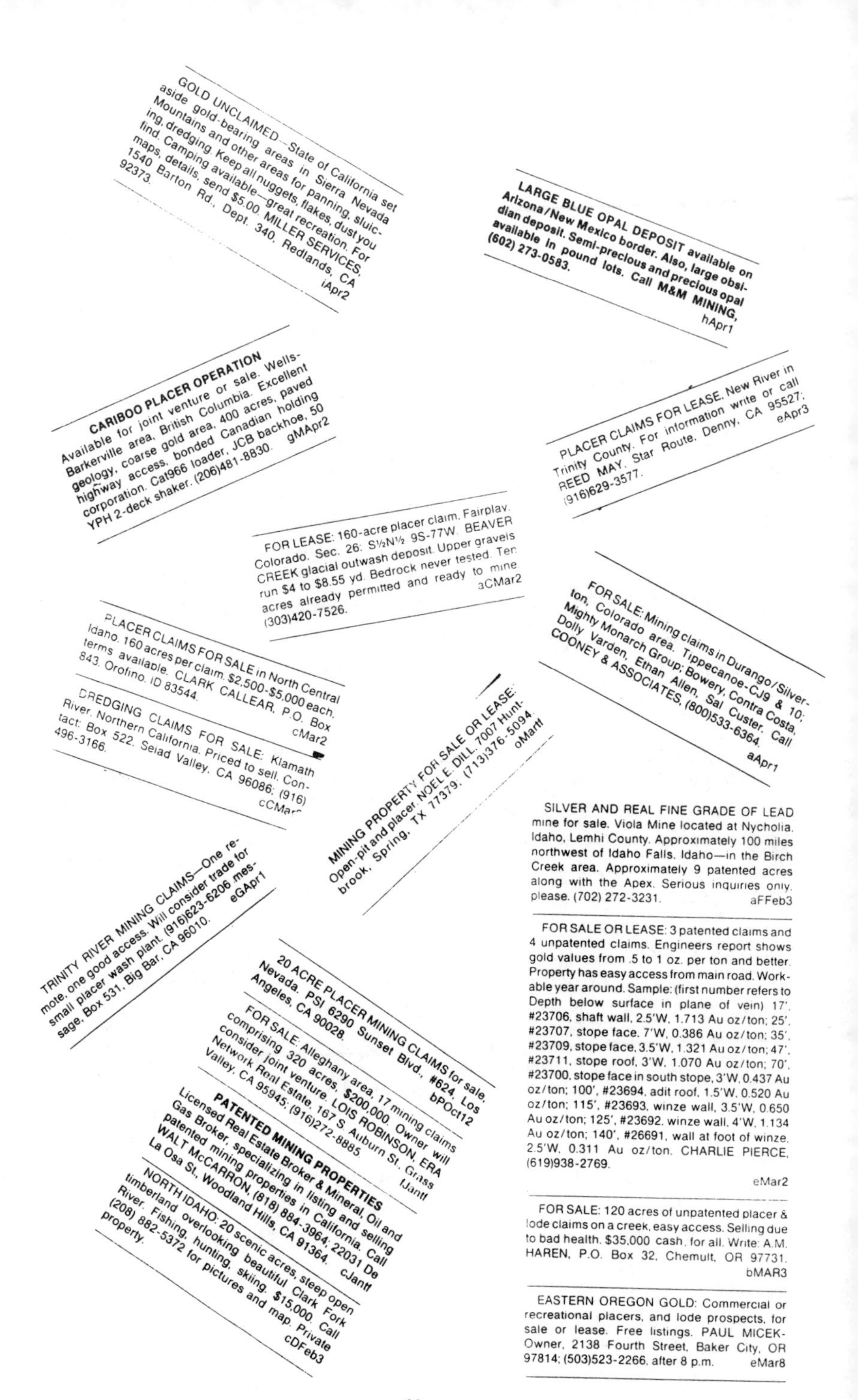

GOLD UNCLAIMED—State of California set aside gold-bearing areas in Sierra Nevada Mountains and other areas for panning, sluicing, dredging. Keep all nuggets, flakes, dust you find. Camping available—great recreation. For maps, details, send $5.00. MILLER SERVICES, 1540 Barton Rd., Dept. 340, Redlands, CA 92373. iApr2

LARGE BLUE OPAL DEPOSIT available on Arizona/New Mexico border. Also, large obsidian deposit. Semi-precious and precious opal available in pound lots. Call M&M MINING, (602) 273-0583. hApr1

CARIBOO PLACER OPERATION Available for joint venture or sale. Wells-Barkerville area, British Columbia. Excellent geology, coarse gold area, 400 acres, paved highway access, bonded Canadian holding corporation. Cat966 loader, JCB backhoe, 50 YPH 2-deck shaker. (206)481-8830. gMApr2

PLACER CLAIMS FOR LEASE. New River in Trinity County. For information write or call REED MAY, Star Route, Denny, CA 95527; (916)629-3577. eApr3

FOR LEASE: 160-acre placer claim, Fairplay, Colorado. Sec. 26: S½N½ 9S-77W. BEAVER CREEK glacial outwash deposit. Upper gravels run $4 to $8.55 yd. Bedrock never tested. Ten acres already permitted and ready to mine. (303)420-7526. aCMar2

FOR SALE: Mining claims in Durango/Silverton, Colorado area. Tippecanoe-CJ9 & 10; Mighty Monarch Group; Bowery, Contra Costa, Dolly Varden, Ethan Allen, Sal Custer. Call COONEY & ASSOCIATES, (800)533-6364. aApr1

PLACER CLAIMS FOR SALE in North Central Idaho. 160 acres per claim. $2,500-$5,000 each, terms available. CLARK CALLEAR, P.O. Box 843, Orofino, ID 83544. cMar2

DREDGING CLAIMS FOR SALE: Klamath River, Northern California. Priced to sell. Contact: Box 522, Seiad Valley, CA 96086; (916) 496-3166. cCMar

MINING PROPERTY FOR SALE OR LEASE: Open-pit and placer. NOEL E. DILL, 7007 Huntbrook, Spring, TX 77379; (713)376-5094. oMar1

SILVER AND REAL FINE GRADE OF LEAD mine for sale. Viola Mine located at Nycholia, Idaho, Lemhi County. Approximately 100 miles northwest of Idaho Falls, Idaho—in the Birch Creek area. Approximately 9 patented acres along with the Apex. Serious inquiries only, please. (702) 272-3231. aFFeb3

TRINITY RIVER MINING CLAIMS—One remote, one good access. Will consider trade for small placer wash plant. (916)623-6206 message. Box 531, Big Bar, CA 96010. eGApr1

20 ACRE PLACER MINING CLAIMS for sale, Nevada. PSI 6290 Sunset Blvd., #624, Los Angeles, CA 90028. bPOct12

FOR SALE: Allegheny area. 17 mining claims comprising 320 acres, $200,000. Owner will consider joint venture. LOIS ROBINSON, ERA Network Real Estate, 167 S. Auburn St., Grass Valley, CA 95945; (916)272-8885. fJan1

PATENTED MINING PROPERTIES Licensed Real Estate Broker & Mineral, Oil and Gas Broker, specializing in listing and selling patented mining properties in California. Call WALT McCARRON, (818) 884-3964; 22031 De La Osa St., Woodland Hills, CA 91364. cJan1

FOR SALE OR LEASE: 3 patented claims and 4 unpatented claims. Engineers report shows gold values from .5 to 1 oz. per ton and better. Property has easy access from main road. Workable year around. Sample: (first number refers to Depth below surface in plane of vein) 17', #23706, shaft wall, 2.5'W, 1.713 Au oz/ton; 25', #23707, stope face, 7'W, 0.386 Au oz/ton; 35', #23709, stope face, 3.5'W, 1.321 Au oz/ton; 47', #23711, stope roof, 3'W, 1.070 Au oz/ton; 70', #23700, stope face in south stope, 3'W, 0.437 Au oz/ton; 100', #23694, adit roof, 1.5'W, 0.520 Au oz/ton; 115', #23693, winze wall, 3.5'W, 0.650 Au oz/ton; 125', #23692, winze wall, 4'W, 1.134 Au oz/ton; 140', #26691, wall at foot of winze, 2.5'W, 0.311 Au oz/ton. CHARLIE PIERCE, (619)938-2769. eMar2

NORTH IDAHO: 20 scenic acres, steep open timberland overlooking beautiful Clark Fork River. Fishing, hunting, skiing. $15,000. Call (208) 882-5372 for pictures and map. Private property. cDFeb3

FOR SALE: 120 acres of unpatented placer & lode claims on a creek, easy access. Selling due to bad health. $35,000 cash, for all. Write: A.M. HAREN, P.O. Box 32, Chemult, OR 97731. bMAR3

EASTERN OREGON GOLD: Commercial or recreational placers, and lode prospects, for sale or lease. Free listings. PAUL MICEK-Owner, 2138 Fourth Street, Baker City, OR 97814; (503)523-2266, after 8 p.m. eMar8

You can see that some mine claims may be bought for an asking price of perhaps $5,000 and some command amazingly high prices depending largely on how well they are established as potential large mine opportunities. In this business you must immediately appreciate that such a property is worth what you can get for it and prices can jump around wildly, depending on the seller's financial condition and his idea of mineral value.

One can say, in general, that the first asking price in most instances will be considerably off what the seller will probably take. Furthermore, barter often goes a long way. Trading property, real or personal, for mining claims is very common and, if you take your time, you will learn that if you can be perceived as a real, competent buyer who is not going to move too fast, first asking prices can vanish faster than fog in the morning.

Time is critical in dealing for mine claims. Try to find out how long the claim has been for sale and why. Spring seems to bring out droves of silly buyers who will pay more than a claim is worth, so avoid competing with them. You have to take time, season, and weather into careful consideration with respect to access. How about snow and mud or dust? You will always hear the best that can be said, so take time enough to see the claim under adverse weather conditions if that is an item.

It is also more important than usual to know something about the neighbors. They could be growing pot, or just be unsavory characters. On the other hand and, more ordinarily, they may be about as fine as you can find. Our neighbors on both sides of the Deep Moon Mine have never given us any trouble and, within a mile or so, there are people who come up and snipe out a few ounces of gold each year to use for Christmas presents while they enjoy the best in family fun camping on their very own spot with a nice picnic table and excellent camp site with beautiful spring water and, almost always, a few eight inch mountain trout ready for frying at the drop of a hook. On the next page there is a picture of their place.

Real property, that is ordinary deeded land, inside a National Forest is either long since taken over or very hard to find at any price but it is possible to find a cabin site or a few acres from time to time and the way to establish this as a yardstick for comparison to the value of a mine claim is to go to the real estate offices in the area of your choice and see what they have to offer and how much the seller is asking.

This camp site, well over a hundred years old, is still enjoyed by our neighbors just south of the Deep Moon Mine.

A mine claim will vary in asking price depending largely on how certain the owner may be of the amount of valuable mineral there is within the claim and how difficult it may be to extract it. Since we assume here that you are basically looking for a "prospect" where you can come to the property and simply explore at first, you can limit yourself to properties that are essentially unproved and, consequently, are not intrinsically valuable as mineral holdings.

Do not expect that information on the average mine claim will be presented in any orderly way; and by the same token, do not expect it to be correct in any respect. Most owners will present what information they do have in almost completely unintelligible form. This writer's experience is that most owners could do better but they either don't know how or they actually want the facts, such as they are, to be obscure. As you sit there in the local coffee shop listening to the enthusiastic seller while he draws on the napkin or brings out a faded copy of an assay of some rock long since lost, and not necessarily associated with the property you are trying to understand, don't ever make the mistake of thinking the property is better than he thinks it is.

This writer has trapped sellers many times who offer an assay as some kind of proof of value by either saying, "Let's go out to your place and duplicate this finding," or by pointing out some basic discrepancy in the paperwork. The sellers, nine out of ten, were unable to come close to the original representation, or innocently "discovered" they had made a mistake and produced an assay from a different place, saying, "Sorry about that."

We can close this chapter with the idea that you must be very cautious, but do not let your caution show. The smarter you think you are, the more likely you are to be taken for a sucker. You must remember that the claim owner/seller is on his own turf and you are the "flatlander" he has been waiting for all winter.

This having been said, we repeat that there are lots and lots of fine people in the country who will treat you right as long as you appear to be the kind of person they wouldn't mind having around.

They do not want you messing up their land or their way of life and they are smart enough to know that you may be able to add a buck or two to the community in one way or another. You will be buying gas, groceries, general supplies, hardware, and other things. You may even need some help of one kind or another lifting things or finding things or constructing things.

The folks in the small town where you go looking around will be very much aware of your presence long before you realize it, and they will be watching to see who you take up with.

You will soon meet any number of individuals ready to give you advice. Some of it will be worthwhile, but most of it will be wrong for several good reasons. You will have to use your judgment here, but do not turn anyone down at first. Let them show their hand. It is probably sound to quote the old saying in the west:

<u>Empty wagons make the most noise</u>.

7. *PRACTICAL CONSIDERATIONS in the Field*

Governed by your personal experience, you may or may not need to follow us closely through this chapter. We have no intention to "write down" to you but, on the other hand, we still remember from a long time ago an incident when we might have done things a bit differently.

* * * * * * * *

It was in late May or early June of 1951. A De Havilland Beaver bush plane had just delivered me to Fort McMurray on the Athabaska River in northern Alberta, Canada. I was Chief Geophysicist for a large consortium of American and Canadian oil companies interested in developing the tar sands in that area, and we had an exploratory program unfolding for that summer in that large, desolate country half way between the American border and the Arctic Circle.

After supper in the local Chinese cafe - this was a very small, outpost type of town - I walked down to the dock where our barges were located to look around. There were five young men making last minute preparations to launch three canoes into the rapid flowing river. In late spring the water is dangerously high with the runoff from winter rushing toward the Beaufort Sea, a thousand miles away to the north.

I noted several basic mistakes they appeared to be making in securing their loads and fastening track lines (emergency ropes) in the canoes, and I felt compelled to make some suggestions. But, they turned me down, saying something like, "Thanks, old man, we know what we're doing." They were geology students from a big Ivy League college back east, headed for Lake Athabaska, two hundred miles downstream.

"Suit yourself," I said looking across that big dirty brown rush of water a quarter of a mile to the other side. Occasional stumps or old logs plunging past told me the river was running at least ten or twelve miles an hour, and it was growing bigger and tougher with every mile downstream from there. Just around the bend there was a stretch of rapids for a good half mile. They shoved off, and I watched them seem to fly away as soon as they paddled out from the shore.

A few minutes after they disappeared from sight I saw a bush plane circle low in the sky a few miles downstream and then head for our airstrip outside of town. Shortly, the pilot and three men came very fast in a jeep telling me they had seen some canoes overturned at the foot of the rapids not more than three miles from town, drifting in an eddy.

In the next two days of searching, all anyone found was the partially destroyed canoes, some bits of baggage, but nothing more.

* * * * * * * *

We do not know of any country in the western United States that is quite as dangerous as the far north, but there are lots of places not more than a few miles from some of our big cities that can and should command our respect, just the same. So, at the expense of hearing someone tell us to mind our own business, we shall recite a few principles just to relieve our conscience if nothing else:

1. Always have the basics with you: a good light ax, 100 feet of light rope (1/4" manila or nylon will do), a good flashlight, matches, a sharp knife, and a small ground sheet.

2. Tell someone where you are going, when you plan to return, and what you would like them to do if you do not show up on time.

If you plan to be very far away or gone for more than a few hours, you should have a sleeping bag or blanket, basic extra clothes, and some extra emergency type food and water with you.

It is very unwise to go into the back country alone. Some do, we know. There still are hermits in the hills, even now. That might be a dramatic term to use, but it is a fact that a good number of men and an occasional woman do live alone back in the wilderness. Many of them around here are Viet Nam veterans, although their numbers seem to be diminishing.

Depending on your method of transportation, there are other considerations. If you are using a car or truck, think about how far you expect to be from civilization. It is one thing if you do not expect to be more than five miles or so from a good, well traveled road. But, it is quite another thing if you expect to be twenty miles or more off the beaten track, or in rough country like the desert or big mountains.

You should check your vehicle carefully before going very far into the back country - battery, cooling system, belts, tires, oil for the engine and other lubricants, even the windshield wipers. Naturally, you want a spare tire and the means to change it.

We always carry a tow chain or cable and a set of jumper wires to hook a good battery to a dead one if necessary. You know your own circumstances and can refine these ideas to suite your specific needs.

This is a photo of our neighbor, Don Buckman, standing beside the notification sign for the Deep Moon Gold Mine on the forest road along the Downie River. Don lives by himself. He has friends who come up from time to time and, especially, in the summer when they work their mine claim, but he spends a good deal of time alone. We should note that there are others along the road, and he is only four miles from town where he shows up at least once a week, and has friends who would be aware if he had not been seen for a day or so.

Here are just a few more general ideas on travel in back country when you expect to be away more than a day:

1. Take enough food for two or three days more than you expect to need, especially emergency type items that will keep under severe conditions of cold, wet, heat, or sun.

2. Never start out late in the day. As a rule, it is wise to plan to stop traveling early enough to get a good campsite set up.

   When we are working with pack horses, we are up and away by five or six in the morning; then we take a good hour to stop and rest around eleven or so. We pull up not later than four in the evening except in late fall when we stop sooner than that.

   With dog sleds, depending on the cold or snow conditions, we try to be on the trail by 8-9:00 a.m., depending on light conditions, and go straight through until about 3:00 p.m., eating a planned lunch on the trail, feeding the dogs at the same time. Winter days are so very short.

   Exploring with canoes is something else. Often, because it is summer and there may be heavy concentrations of mosquitos or flies near the water, we may even cook right in the canoes, lining the bow with a couple of inches of clay and lighting a small fire down out of any wind. Overall timing can depend as much on when the bugs are out as anything else.

Obviously you have the most flexibility when you are traveling with vehicles, but it is still wise to use the early part of the day, and never run until almost dark before settling into a strange place.

Preparation for any trip of more than a few hours is critical to your success. We have check lists that include: a) the general items that will be necessary; b) the personal items; c) the special items such as survey tools, instruments, maps, etc.; and d) the food items including specific accounting for water (where we will get it along the way if we do not take it all with us). We have not mentioned three items: a good shovel, a camera, and a notebook for a diary. Keeping a diary of any trip is basic. You should note location and weather each day routinely to start with.

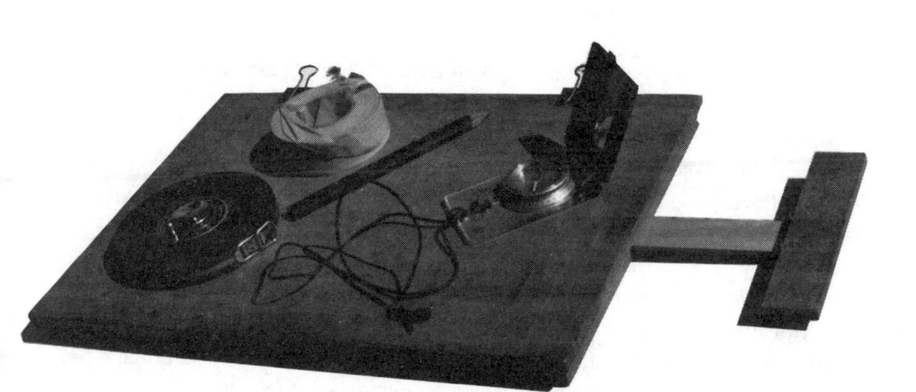

The author has used this small sketching table for over forty years. Notice it has an attached T-square which is very useful. The compass is Swedish. A Brunton compass is "de rigueur." The surveyor's tape and heavy carpenter's pencil are expendable.

In the old days we always thought it was a good idea to have the simplest possible objects and materials on an expedition of any kind. Two items were thought to be indispensable, a few square feet of cloth and canvas and sewing equipment.

If we stove a hole in a canoe, we could whittle and bend some ribs out of green wood and patch it with the canvas, making the thing watertight with some sap from a tree or pitch or something. We would fix a tent the same way if something fell on it or it was torn for any reason. We could even fix an item of clothing or make a sling or even a splint for a broken bone. This writer has traveled for several weeks in the north after breaking both bones in his forearm, and setting them right there in the wilderness with no benefit of the niceties of the medical profession at all - just unflinching help from a tough minded, common sense companion guide.

Today, with all the synthetic materials around us, things are forever different. Most canoes, for example, are made of some kind of tough fiberglass or other petroleum based material which would be very hard if not impossible to mend outside a specialty shop. On the other hand, they are much tougher than an old fashioned canvas canoe, so it may balance out. Similarly, the tents are made of a very light polyester or something. Of course, they can be repaired in the field and, frequently, come with their own repair kits.

Sometimes, surrounded with many camp items including clothing from jackets to socks, backpacks, and lots of other things, we have sat around a campfire contemplating how some belligerent zealots, in the name of animal rights, clean air, or environmentalism, think they can travel through the wilderness depending, like us, on the wonders of the oil business, not to mention the fuel used to get around at all while, at the same time, trying to bring that big industry to a grinding halt if they can or, in another case, denying the National Forest to a sheepherder or a cattleman.

A word about tents -

The shape of a tent can be very important. We see many neat little efforts in the sporting goods stores today that appear to be very suitable for people in a campground just out for a good time while setting up their outfit within ten feet of the big van they came in, loaded with thermos containers and large ice chests and Coleman lanterns. There is absolutely nothing wrong with all this, if that is what turns you on, but if you are seriously out to find that hideaway we are talking about, you will probably have to stay out longer than a week-end and often find yourself further out in the back country, not in a planned government campsite.

You can get caught in a torrential rainstorm, even in the desert, where you find yourself wet through. It is possible, in this situation, that you can simply wait for the sun to come out and spread everything over some bushes to dry. But, in a great deal of the beautiful wooded country, it might not be so easy to dry out. Here a good old fashioned wall tent with a little height to it can be a Godsend. You can hang the wet things up inside, even while it is still raining and begin to salvage critical items, especially food or clothes.

A tepee is something else! The big problem today is the poles. In the old days, all over the west, especially in Canada, you could find Indian campgrounds with tepee poles neatly stacked for that purpose. It was thought to be proper for any of us traveling through to cut a few new ones now and then or to leave some other useful item for general use such as a portable wood stove for heat or cooking.

The author mapping a 20,000 acre petroleum concession on the North Pine River in northern British Columbia. Note the dense distribution of young dead trees in the background. This is called brulé country - extremely hard to move in. Four days after this picture was taken, the party, having moved on to a muskeg, was taken by surprise when seven inches of snow fell and they almost perished. August 1953.

Exploring for that hideaway can be loads of fun, but things can take a turn for the worst, and always at an inconvenient time. Being cold and wet with no immediate salvation can do you in.

With a wall tent and good headroom you can have a small, folding wood stove with a 2" pipe stack and elbow so the smoke and fumes can escape out from the side, not the top of the tent. In country stores and some outfitter stores, you can buy such a stove and a fireproof insert that fits the stove-pipe which can be sewn into the tent wall at an appropriate place. Serious wilderness travelers use this kind of cover to survive unpredictable storms.

A tepee is without question the most romantic of all tents. It is built to have an ordinary campfire inside surrounded by a ring of rocks or dirt. With the fire placed back of the center, away from the door, the heat rises following an S-shaped path before escaping out through the flaps, and the tent can be surprisingly warm even in sub zero weather. The author lived in such a tepee on the open desert southwest of Grand Canyon for a year. Knowing he was going to stay a while, he installed a flagstone floor with a raised fireplace and grill cooking area, with extra lines secured to stakes outside to insure against being blown away in a location where you could see ten miles in any direction. He and his companion stayed very comfortably through several severe winter storms and winds.

You will almost certainly choose a less cumbersome shelter than a tepee when you first start looking around in the countryside but it is a fine place to live during that period when you have found your hideaway and you are having fun looking for or developing your gold claim before you build that dream cabin.

This is a picture of the Athabaska River taken at low water in the early fall. The gorge near the left is about a hundred yards wide, and the white line on the far bank indicates that the water in the spring is about eight feet deeper than this. Not exactly the place to learn how to paddle a canoe!

There is something else to think about besides bad weather before we leave the subject of camping and traveling in the wilderness.

In unfamiliar country strangers have been known to find themselves turned around or even lost. Again, we do not wish to offend you, if you are already expert in these things, but the tragic memory of those young men drowned in the far north so long ago compels us to make a few observations here.

1. While you still know exactly where you are, study the country ahead where you intend to go. If you can see, make a firm mental note of the general lay of the land, the high points around or other broadly distinguishing features that you can refer to as you go along. Review your basic orientation with the compass. Make a note of the time of day you are actually ready to take off, so at any time after that you can make a good guess about how far you have gone.

2. As you move along, look back frequently so you have a good idea of what the country will look like when you decide to return. Leave a blaze on a tree if there is any doubt, or take some flagging and use it as needed. Do not try to move too fast. Do not allow yourself to become so absorbed in something else that you fail to pay attention to your location and the general direction of your movement.

It is not necessary to cut a slice out of every tree you pass. If you are in heavily wooded or brushy country where you cannot see very far in any direction it is wise to travel as nearly as possible on compass courses and change directions at specific, recognizable spots, blazing or hanging flagging at and near these turning points. Flagging is preferred over blazing most of the time but small animals, cattle, deer, and even birds have been known to eat this stuff as fast as you can hang it in some areas.

Day or night, travel is not difficult in ordinary country if you can see the sky. There is always the sun or the moon or the stars to guide you but you have to remember that they appear to move through the sky at the rate of about 15 degrees per hour, and you have to take that into account. Difficulties arise when it is cloudy or raining especially when the country is monotonously similar mile after mile as on open prairie or in densely treed areas. In all such situations you are wise to have a big picture of the country in your mind - where any rivers are or other distinguishing topographic features even at a distance; and never go into that kind of country without profound respect for its potential to get the best of you. We are reminded of the old saw about the sailor finding himself on the open ocean in a row boat, "Lord help me. Your sea is so large and my boat is so small."

We have felt a clear obligation to discuss the tough side of travel in the back country, but let's consider the Tahoe National Forest where the author's Deep Moon Gold Mine is located. The map covering this part of the world is about three feet by three feet and covers almost 5,000 square miles. That is a lot of big country, but it is just one national forest of many in the western United States.

The map shows roads: Interstate highways, ordinary paved country roads, all weather gravel roads, four wheel drive roads, trails and hiking paths - thousands of miles of them! - not far apart; so it should not be too easy to get lost. Still people do, and the Sheriff's Search and Rescue Teams organized in communities throughout this country are busy, especially on and after big holidays.

You could be lost and snowed in too! This is the author's little Audi rescued from a bit of Sierra snow. For a while we thought it was lost but we found it outside his door!

The U.S. Forest Service uses thousands of little yellow metal plates in a very interesting way to help people know where they are as they move over these roads and trails. They are called K-Tags or K-Markers. The word is that they may be technically called "Forms" and the "K" derives from some old form number, long since forgotten. Anyway, they have marked on them a set of squares depicting a 36 section township of the standard variety. Each section is numbered, and the idea is that the forest rangers go out on the roads and trails and fix these tags to trees usually with ordinary nails, and pound a nail hole on the line of the appropriate section where that particular path or road crosses the section line. The Township and Range of the tag is etched on it so you have identification of the spot where you are. All you need to do is read the tag, then look at your National Forest map and find out within a hundred feet or so exactly where you are.

This is a K-Tag:
A Forest Service Location Poster.

Note that some stupid, thoughtless individual has fired three bullet holes into it, one above the other. It appears he or she was not a bad shot, but it certainly would be no help to some city person lost on a cold winter night trying to figure this out. Fortunately you can still tell the location by the smaller hole down and to the right on the line between Sec. 26 and Sec. 27.

This is a K-Tag on an actual tree in the forest. sometimes they are a bit hard to find, because of the awkwardness of the location itself.

You can also miss them easily due to the fact that you are watching something else at the time you pass one of them. But, most of the time, they are in the right place and can be very useful.

We probably should at least mention the subject of first aid and supplies. Some people may wish to have a good first aid kit, put together by professionals who undoubtedly know what they are doing, but seem to feel obligated to provide for about every weird catastrophe that you can imagine. It is not uncommon to find a little booklet of fine print in these boxes of medical things which covers all kinds of problems from abdominal wounds to watching for shock, with chemical burns, drowning, epileptic seizures, fractures, frostbite, nose bleeds, and poisoning, among dozens of other scary possibilities. We are especially amused by the instruction: Send for the doctor.

We leave all this entirely up to you with the observation that we do not even carry aspirin tablets.

Let's move on to the day when you have found your mine claim and are ready to do something with it.

The schematic sketches on the following page illustrate a useful type and size of storage building which can be constructed with a minimum of effort and cost; and at the same time be quite secure from vandals or others like that character shooting up the K-Tags who would attempt to take objects from your claim in your absence. It is made with 6" x 8" x 16" concrete blocks. The standard, as you probably know, is an 8" thick block, but this is extra heavy. There is a 4" thick block, but this is not structurally sound. These blocks are placed on a 3" cement pad with a bit of reinforcing wire in it. You top this off with a wood roof with varying pitch depending on snow conditions. The 6′ gable makes storage room above the 2" x 6" cross members spaced on 24" centers. To be fireproof, cover it with composition shingles so it is relatively inconspicuous, not shiny.

With an inside dimension of 10′8" wide, 11′ long, and 6′ high, there is enough space for equipment, hand tools, lawn chairs, table, and even a 6" dredge with a 16hp gas engine, hose, sluice, etc. If you want more headroom, one more tier of blocks will give you 6′8."

The 12′4″ x 20′4″ pad allows for the 11′8″ x 12′0″ o.d. building to have a 4" margin area around it and have an 8 foot extension out from the front of the shed making this much better than having dirt right outside the door. The door may be two simple 4′ x 6′ sheets of 3/4" plywood with appropriate edges in a simple door frame which can be lifted into place and locked, or you can be fancy and have hinges.

We have seen elaborate variations of this basic building. You dig a small square hole inside the door to the right or left and line it with concrete, deep enough to provide cool storage with an insulated wood cover. The building can also be located over your well head if you develop your own water and contain a pump and electric generator.

In that case some people will build a vent into the roof and surround the generator with simple insulated plywood to keep the noise down in camp. Naturally, you can put in shelves. The conveniences you want are up to you but the idea is to make it no larger or higher than absolutely necessary so it will not attract attention; and keep it as simple as possible so it is hard to break into if you expect to be gone for long periods of time.

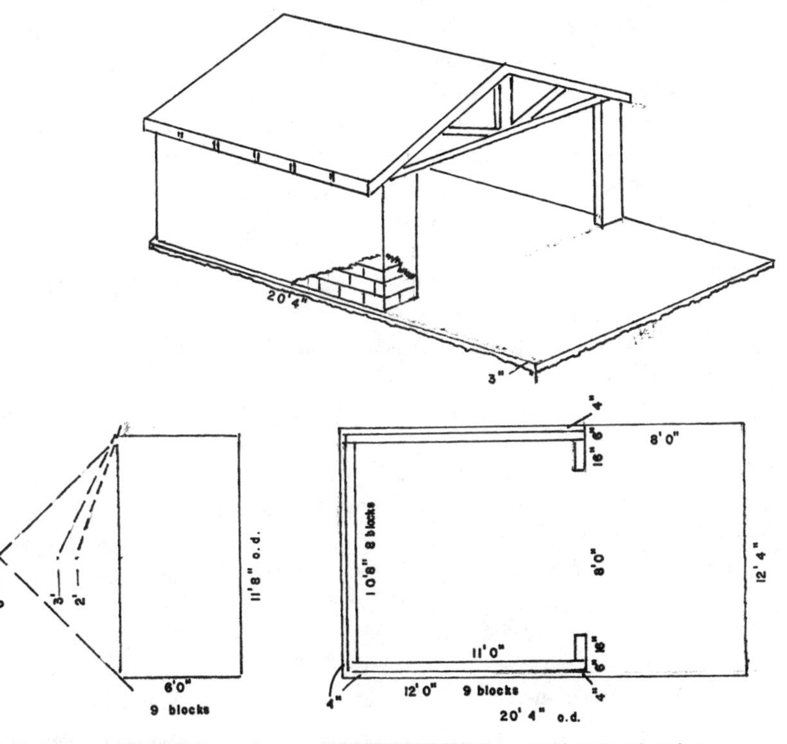

CONCRET BLOCK STORAGE SHED schematic sketches

When you are ready to look for some gold, you should drop by the nearest U.S. Forest Service office to find out what permits you may need. Most people know about campfire permits. They are issued each season by the hundreds of thousands all over the country. But, there are specialized permits required for gold miners.

FEE: $26.25 RESIDENT
FEE: $104.75 NONRESIDENT
(Submit with application)

STATE OF CALIFORNIA - THE RESOURCES AGENCY
DEPARTMENT OF FISH AND GAME

1991 STANDARD DREDGE PERMIT APPLICATION

TO OPERATE VACUUM OR SUCTION DREDGE

I hereby make application for a standard dredge permit to use a vacuum or suction dredge.

DATE OF BIRTH ______/______/______
Month Day Year

NAME ____________________

HGT ________ WGT ________

ADDRESS ____________________

COLOR OF HAIR ________ COLOR OF EYES ________ SEX ______

CITY/STATE/ZIP ____________________

TELEPHONE (   ) ____________________

____
DRIVER'S LICENSE OR SOCIAL SECURITY NO. (OPTIONAL)

I have resided in California continuously for the last 6 months. *(A resident is defined as any person who has resided continuously in California for six months or more immediately before the date of application for a license). Check one:* **YES ( ) NO ( )**

TYPE OF OPERATION *(check one):* **GOLD MINING ( )** **SAND & GRAVEL ( )** **OTHER ( )**

Explain "Other" if checked: ____________________

I hereby certify that I have read the provisions of the California Fish and Game Code, Section 5653 (see reverse side for copy of this section) and that I understand and agree to be bound by all the terms set forth in the permit issued pursuant to the above named section.

I hereby certify that all information contained on this application and/or submitted to meet the requirements for renewal of subject permit(s) is true and correct. I understand that, in the event that this information is found to be untrue or incorrect, the permit issued will be considered invalid and must be surrendered where purchased and that I will be subject to criminal prosecution.

Signature of Applicant ____________________ Date ____________

When permit stamp has been affixed below, the applicant is authorized to operate a vacuum or suction dredge with intake diameter of 8 inches or less and in waters open to dredging in accordance with the attached list of open and/or closed waters. The standard permit authorizes the discharge of dredged material below the ordinary high water elevation. No dredged material shall be placed into wetlands adjacent to waters of the State. Dredged material is limited to native streambed materials processed through the dredge equipment and rocks moved by hand during its operation. (Fish and Game Code, Section 5653).

Nothing in this permit shall authorize the permittee to trespass on privately owned land, or to use a dredge in waters passing over private lands without permission of the landowner. The listing of waters open to dreding does not mean that such waters are open to the public. The permittee shall conform to all applicable federal, state and local statutes and ordinances. Suction or vacuum dredges shall not be used where dredging is prohibited by ordinances, statutes or regulations adopted pursuant thereto. This permit does not authorize dredging in any national forest, national park, state park system unit, county park, municipal park or other such area in which dredging is prohibited by the agency in control of such areas. **Permittee shall file a notice of intent with the U. S. Forest Service for approval if dredging is to be done on National Forest Service lands.**

FOR DEPARTMENT OF FISH AND GAME USE ONLY

Resident Dredge Stamp
Expires December 31, 1991
No. ____________
Date Issued ____________

Nonresident Dredge Stamp
Expires December 31, 1991
No. ____________
Date Issued ____________

____________________ ____________ ____________
Signature of Department Representative *Title* *DFG Office*

White - Licensee Yellow - Issuing Office FG 1385 (Rev. 11/90)

Gold can be counted on, because it is so heavy, to sink to the lowest place it can find, following flowing water or simply the influence of gravity. Under rocks or very large boulders is well known to be an excellent place to look, so it is not uncommon to face the problem of moving or at least rolling over a boulder eight feet through weighing several tons. In this case you must have a special winching permit.

Your best bet is to go to the U.S. Forest Service first, and tell them exactly what you want to do and where. There is no point in trying to be the least big cozy with these people. Sooner or later a ranger will be drifting by, just looking around; and if he finds you in compliance, you have made valuable "brownie points." If not, the authorities will simply begin to watch you more closely.

The Forest Ranger will tell you, as soon as he hears what you are planning, who else you may have to deal with, but it is certain that you will be dealing with the Fish & Game people if you are in the state of California. They are the people who issue the dredging license and the winching license to all gold miners, big and small.

8. *GOLD, SWEAT, LUCK and JOY*

Developing your claim and enjoying it will probably require many years of effort while having about as much good, wholesome, truly re-creating time, which we all call "recreation," as you can imagine.

You can play hermit, and do it more or less by yourself, or you may have ten people around - with enough to do to keep everyone busy clearing places, building things, hunting for gold, and taking time out for a nice long siesta in the afternoon sun, or a swim, or just a cool one. Your place can also serve as a kind of headquarters for exploring the country around you. The possibilities are virtually unlimited.

We can only imagine what you may have found that turns you on.

Is it a real hideaway in a glen back near the headwaters of a little stream where you can only get in by a precarious path with a llama - with damp moss all around and the possibility of real gold in every niche and crack, under the tree roots, between the rocks that part the tiny stream of water coming from a spring above your camp?

Or is it a windswept crevice on the side of a mountain with a big dark streak of galena carrying a hundred ounces of silver to the ton, and you will have to construct an ore chute a mile long down to a shed where you can bag the stuff and send it to a refinery five hundred miles away?

Or have you joined several other people where you work to develop a five hundred acre alluvial fan out in the desert; and you are about to move in a backhoe to make test holes twenty feet deep about two hundred feet apart so you can peddle the claims to a company that already operates in the area? Hardly a hideaway, but you stumbled on to this place when some old guy promoted you into looking at it based on one lonesome mole hole he dug more or less by hand where he pulled out enough gold to convince you it was a good bet.

Or did you get sidetracked looking for that old Spanish treasure shown on the faded map the guy sold you for a hundred dollars - the one he said was given to him by an old German hermit as he was lying on his death bed, barely able to speak? There was that one complication (there always is). But you and your friend think you can figure that out when you climb to the place on the side of the mountain where the arrow is carved in the face of the cliff.

We remember an admonition first heard in an old cabin at 50-Mile on the Yukon River while we were waiting out an early snow storm:

"As you wander on through life, brother -
No matter what your goal,
Keep your eye upon the doughnut,
And not upon the hole."

The prospect of gold can do strange things to a man's mind.

Your best bet is probably to hold to the original objective of this little book - just find that nice hideaway where you may pick up a few small chunks of the precious stuff, and settle for some real peace of mind away from the madding crowds as you simply sit back and enjoy the wonders of nature and take a little time to contemplate life.

You could be dredging like the guy in this picture, run onto a hole in a northern California river and come up with a prize like the one his wife is holding on the opposite page. Talk about a high!

These are pictures of real people, and that is real gold! Just a few shots of this size and you could pay off the mortgage.

Once hit with the gold bug though, people rarely just walk away and cash in to do something logical.

Our neighbor hit it better than this a few years ago. He and his wife were smart. He still gets good gold out of the claim where he found a real bonanza which made a whole lot of money.

Today they own a nice home here in Downieville, another one in Arizona, a few rental properties and one of the most profitable stores in town, where they sell gold from here, Australia, and lots of other places as well as about a thousand other items.

This is a picture of our friend in the wet suit and his wife (the one holding the pot of gold) having a good time at their campsite near their mining operation. They run a nice travel agency today, but you may be sure they will never forget those good old days on the river.

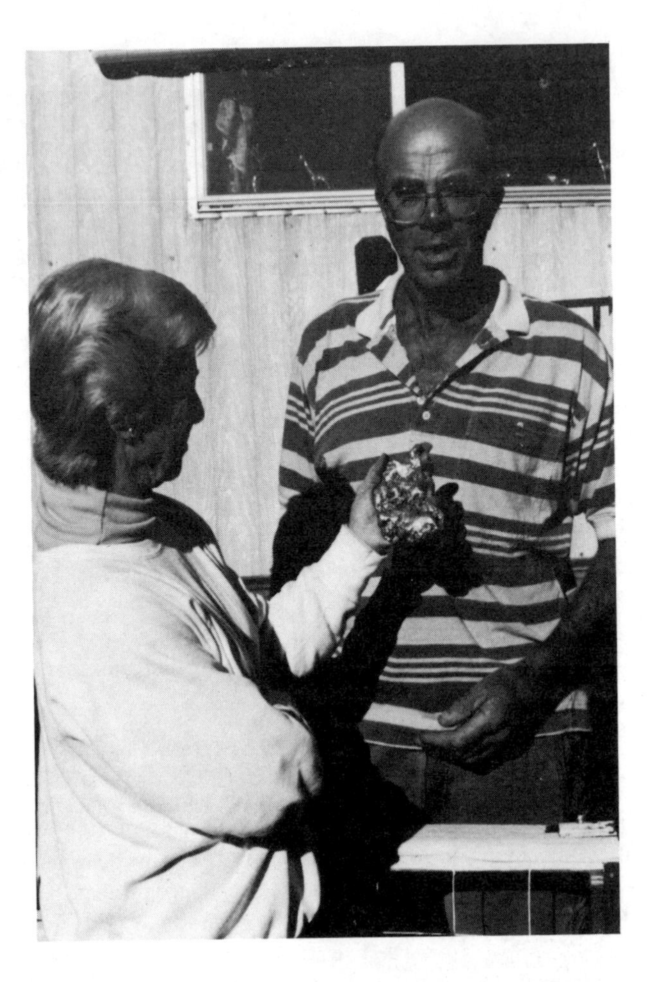

This is a picture of our neighbor who found the bonanza on his claim. Our good friend, Peggy, is admiring a 40-ounce nugget he sold that brought in about $25,000. Specimens like this often bring as much as two times the spot price of regular gold. It depends on who wants it and the intrinsic evaluation depending on how it is judged as an art object. A slug, a big nugget with no "class" may go for no more than spot, and sometimes less, because these are not pure gold. Nuggets will generally run between 75% and 90% pure, and the impurities can range from iron, copper, silver, elements of the platinum group or other chemical elements which cause interesting variations in color and overall attractiveness.

The guys in the picture below were mostly Viet Nam vets with a homebuilt wash plant that was probably losing almost as much gold as they found. The equipment was old and not very reliable. But rumor had it that seven of them with their wives or girl friends were able to live more or less well on this operation for a while. They might have done better but they had a love of automatic weapons and had the habit of firing off at least a hundred dollars worth of lead into the air around their claims every afternoon after work "just for the hell of it," and to keep the people nearby a bit scared of them so they would not have any trouble being ripped off.

Each of these men had a coffee can stashed under a tree stump for their gold. At the end of the day they told this writer that they spread out an oil cloth on an old table and divided the day's gold into seven piles as near equal as the eye would allow. They had a regular order each time, so one day one of them would do the dividing and the next behind him got first choice and so on. The next day the man behind did the dividing and the one behind him had first choice. This went on through seven rounds and then they started over again, figuring that things would even out over a period of time. It was a pretty cool operation while it lasted and, whenever one of them needed something, he just dug into his can and took some gold to town.

Beginning with those folks and their 5" to 8" dredges working in open rivers, next the characters working the somewhat larger fixed plant back in the wilderness, we show here an operation out in open country that can probably handle more gravel in an hour than our happy friends can do in a week.

After half a lifetime involved in extractive mineral enterprises, this writer has concluded that the very small efforts usually begin with about 5% intellect and 95% emotion. That is to say the folks involved simply do not know what they are doing, but they are having a very good time. As they learn and are exposed to larger opportunities or experience, the intellect/emotion factor shifts through the very dangerous 50/50 range to 95% intellect and 5% emotion where the protagonists -operators of the giant mines - are characterized by having, as we say, "ice water in their veins." They go strictly by the numbers, and there is absolutely none of the old Robert Service human element in their judgment at all. We know nothing specific about the operation depicted here. We don't even know where it is, but we will evaluate it as about a 40/60 deal, very risky, because judgments are probably 60% emotionally based with only 40% founded in actual facts.

A hideaway is a hideaway and a gravel pit is a gravel pit. A few nuggets, found every once in a while is one thing, but a steady stream of fine stuff, grey with residue from lethal chemicals is something else. The temptations to turn a hideaway into a gravel pit can be overpowering, but you had better know what you are doing because you are much more likely to lose your shirt than anything else.

A couple of stanzas from the famous "Bard of the Yukon," Robert Service, say it better than we ever could:

From "The Spell of the Yukon"

I wanted the gold, and I sought it:
I scrabbled and mucked like a slave,
Was it famine or scurvy - I fought it;
I hurled my youth into a grave,
I wanted the gold, and I got it -
Came out with a fortune last fall, -
Yet somehow life's not what I thought it,
And somehow the gold isn't all.

From "The Prospector"

I strolled up old Bonanza, where I staked in ninety-eight,
A-purpose to revisit the old claim.
I kept thinking mighty sadly of the funny ways of Fate,
And the lads who once were with me in the game.
Poor boys, they're down-and-outers, and there's scarcely one today
Can show a dozen colors in his poke;
And me, I'm still prospecting, old and battered, gaunt and gray
And I'm looking for a grub-stake, and I'm broke.

We really get the message from the *last stanza of "The Spell"* -

There's gold and it's haunting and haunting:
It's luring me on as of old:
Yet it isn't the gold that I'm wanting
So much as just finding the gold.
It's the great big, broad land 'way up yonder,
It's the forests where silence has lease;
It's the beauty that thrills me with wonder,
It's the stillness that fills me with peace.

TEMPTATION!

The white object at the top of this picture is a one-foot long engineer scale to give you an idea of size. This is about one hundred ounces of gold sorted out in sizes.

A sight like this can turn your head around, and upset the normal balance between your good sense and your potential for excitement.

To put it another way - gold can drive you crazzzzzy!

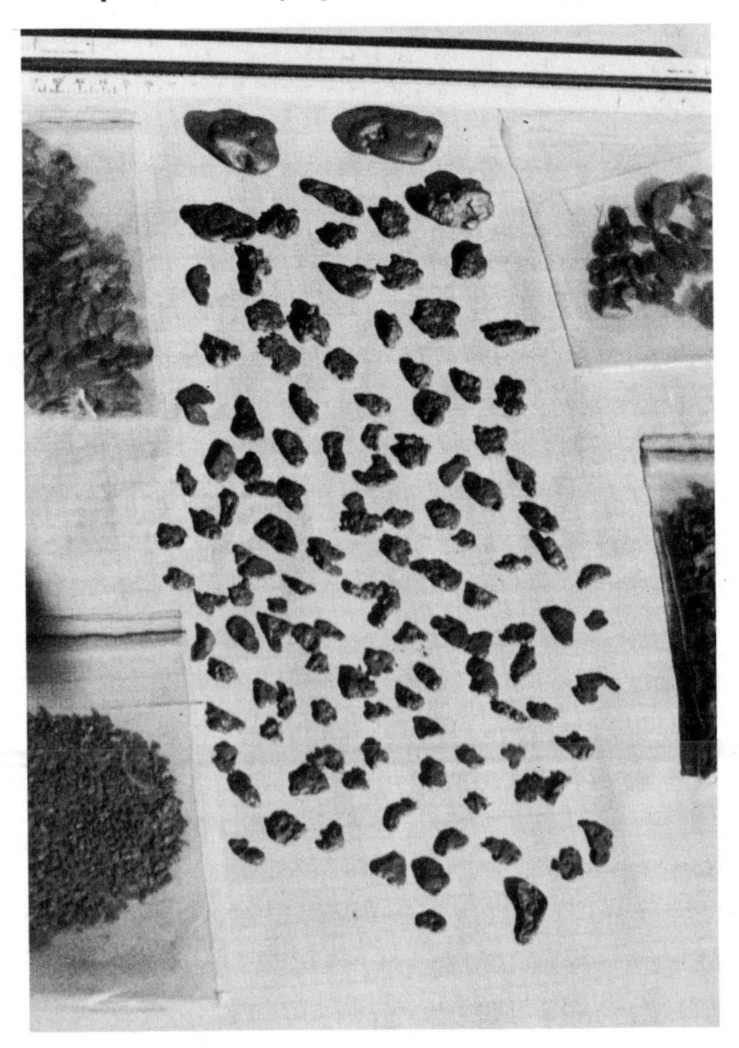

Long ago when television was new, we were invited to do a half-hour show for a New York City TV station on commercial exploring in far north. At that time we prepared a photo essay to answer the question: What was it like to be an explorer in the Far North? Here is an excerpt.

Well you sat around mostly in the winter unless there was a special job to do.

You studied photos and maps and listened to tall stories.

But when spring came you headed for that special bend in the river or the mountain you saw last summer.

There it is. See that black area near the top. Is that just a shadow — or maybe it's an outcrop.

We're in the secret part of the Cassiar in northern British Columbia, and in this crisp spring air that might be six miles.

Closer — and it still looks interesting. But we'd better go back and move the base camp. Men have been known to get too enthusiastic and never had a chance in a mountain storm without being ready.

Here at last. This is "Rattlesnake Bill" Puritch. He got that name for selling snakes to the Dookabores for their ceremonies when he wasn't out prospecting.

The rock is heavy.

The samples check out! High grade galina — lead, silver, and zinc.

So we send for the magnetometer to see how big it is.

Here we unload the geophysical gear.

Nothing can stop us now. Only problem we have is some minor legal problem among the Partners in the city.

Hear Tellington Associates stock jumped.

It's at least 5′ wide and 226′ long. Wow! $1,000,000.

A Tellington Associates Ltd. bulldozer arrives.

We're going into the mining business —

That beautiful, big 9,000′, unnamed mountain is cut. It will never be the same again.

I don't feel so good. Bill and I and the surveyor got drunk on Hudson Bay 180 proof rum.

Having discovered a valuable mineral site, the next thing to do is develop it. This is often done in stages, beginning with a process of proving up the deposit or blocking it out by one practical method or another, to find out what the total extent of the mine may turn out to be.

Here we elected to install a pilot plant operation to dig back to establish the nature of the vein and check the quality as well.

It is a strange and tragic fact of life that success for an explorer simply means limiting the horizon for future individuals to an ant-like mold.

Soon we will be, like ants, crossing this creek with systematic simplicity on our way to work.

Did I say something about ants?

The ore chute grows —

until it is a mile long.

By the sweat
of your brow
shall ye live.

Conquered.

The heart of our mountain is sacked and weighted in this shed.

Wouldn't it be a tremendous achievement if we could make

the whole mountain into money!

Is that 7 tons of rock in those bags — or is it $1,000?

Hydraulic mining at the famous Malakoff Diggins, North Bloomfield, Nevada County, California in the last half of the 19th Century. Fine dirt from this operation floated in the Yuba River System all the way to San Francisco Bay and forced legislation to stop it.

9. *THE COMMERCIAL DIMENSION*

Although our thesis is focused on the early stages of mine claim development - where the fun is - we have to keep in the backs of our minds the idea that taking advantage of the law to record a mine claim at all is serious business, and you might run on to a bonanza type deposit that would demand your "going commercial." In that case you would proceed with some kind of intelligent development and then turn your attention once more to finding that hideaway somewhere else.

The picture on the opposite page shows some old timers working down a big bank of high bench gravel to recover gold from a hillside about twenty miles south of the writer's Deep Moon Mine and ten miles northeast of Nevada City, California, in the late part of the 19th Century. Such an operation required "capitalization."

Very few of the original "49ers" went commercial, but in the years after they opened the country up to more systematic exploitation, there were many mining operations organized all over the western part of North America which required more money than any one man might have or be willing to gamble by himself.

This kind of growth invariably begins with someone making a discovery on the ground of a valuable mineral deposit. That person might be an employee of a large mining company but, over the years, the discoverer, more often than not, has been an individual prospector with barely enough personal resources to live from one week to the next - just like the guy in Robert Service's poem.

The great Noranda mine at Yellowknife, isolated on the north side of Great Slave Lake, latitude N 62.5, a thousand miles north of Edmonton, Alberta, Canada, was found by an intrepid guy who shared an interest in his claims with the bush pilot who flew him in. He sweated and strained, and the bush pilot, according to local stories, became the President of Noranda, one of the largest and richest mines in the world, with several miles of underground tunnels yielding gold and other minerals including radium.

Traditionally, the first non-operating interest in a mine claim was called a "grub-stake," and it is what it sounds like, an interest in the land staked as a mine claim in consideration of providing the prospector with the food he needed. It then expanded to include the price of a shovel, a tent, other simple tools, and even transportation for the prospector - any out-of-pocket type of expense that might be incurred in the original effort.

The next interest, traditionally accepted in the mining business was called a "working interest." This was not a grub-stake. Basically, it meant something quite different. It was equivalent to a "sweat equity" earned by working for the miner who owned the claim. It could also mean a "money equity" earned by providing the prospector or miner with help from a simple grub-stake to providing machinery or supplies. Following up on the notion of work, this was "working capital," but it did not imply any ownership as such, however. Whether it was money or work, the operative notion was "help." If the help quit, the miner had no further obligation whatsoever to the person who left; and similarly, if the individual providing the necessary capital quit putting up money while the job was on-going, he had no further interest of any kind. In the old days the prospectors and miners had no use for "quitters."

Continuing on with the kinds of traditional interests commonly accepted in the mining business, we have an idea that has been around for centuries. This is defined by the word "partner" and carries much connotation based in law and practice. People can become partners by anything from an accident to a long legal document, but the idea is that each can speak for or contract for the other and share as agreed in the common enterprise. There have been refinements of this notion over the years where one person can be a "limited partner," having no say in the operation, but still reserving an interest, while another person can be a "general partner," carrying the responsibility for decisions and acts related to the enterprise. In the mining business you have the idea of working interest or the idea of a partnership. They are very different and mutually exclusive. Many a miner, as the old saying goes, will deal with a person but not want to get in bed with that individual. That is the philosophical difference between a working interest and a partnership.

We come to two more concepts or notions regarding relationships in any enterprise which can become especially important to understand.

We have the idea of perpetuity. What happens if the miner dies? What happens if anyone dies who is involved in a project of any kind? Then we have the idea of liability. Who is responsible, and to what degree are they responsible?

The answer to these problems is normally found in the notion of a company or, more specifically, a corporation. The dictionary defines a corporation as "an association of individuals, created by law, and existing as an entity with powers and liabilities independent of those of its members."

The very first such organization in history was the Hudson's Bay Company formed in 1668 by a French explorer named Radisson, who promoted a number of courtiers at the court of Charles II of England into joining some merchants in London in outfitting and financing an expedition to open fur trade with American Indians and find a route by sea to the Orient. It was formally "chartered" by a grant from the King to Prince Rupert in 1670.

Here we have several fascinating "firsts" in history:

1. Radisson, the promoter, had an interest, but more like a paid employee than anything else. He had the "idea."

2. Prince Rupert and his courtier friends may or may not have put up any money. They were primarily parties to the operation because of their "influence" with the King to get the necessary permission.

3. It was primarily the merchants, a clearly lower class of people, who put up the money and supplies, a kind of secondary group, if you will. They supplied the "material things."

4. The Charter was granted to "these Gentlemen Adventurers trading into Hudson's Bay," and all parties had paper in hand to show what their agreed share in the enterprise would be.

The Hudson's Bay Company became generally known simply as the Great Company, and operated with unique authority and great profit until the middle of the 19th Century, when financiers entered the picture after some powerful politicians of the day conducted a parliamentary investigation inspired by enemies of the company. The Hudson's Bay Company was internally reorganized and, in 1863, the stock of the company was bought up and reissued by an organization called the International Financial Society.

This was essentially accomplished by a "paper transaction" done and agreed to by the parties for mutual benefit as they understood it at the time.

Here we have reproduced a piece of paper - a very simple document issued from some entity called McDame Creek Mines, Ltd. - indicating that a man named William Houslbut is given 300 shares of stock in consideration of "equity" in a mine plus one dollar in cash. This certificate, as you will note, was issued in 1954, - nothing more than a piece of paper (probably inspired by blind faith!) for something presumably quite real. People with various and varying interests in mining activities have traded such things almost like money for whole mines or a piece of heavy equipment or most anything else.

PHONE 28466

McDAME CREEK MINES LTD.

SILVER, LEAD, ZINC

LOWER POST, B.C. :: EDMONTON, ALTA.

Aug 23 1954

Received from William Houslbut

The sum of Equity in [illegible] Plus 1.00 dollar in Cash Dollars

FOR 300 Shares in McDame Creek Mines Ltd Capital Stock

$3000.00 Value

PER [illegible]

REC'D BY

This tells you a lot about simple country people and miners in particular. They really will do things like this, but don't stretch your luck. Many a cheating city slicker has been run off the road or otherwise brought to a kind of back country justice while lawyers try to find their briefcases. Still we must observe that lawyers have a way of tying things up to the point where even judges can throw their hands up in despair; and you can be locked into limbo for years.

Simply try to know what you are doing and state things as clearly and simply as possible as you go about converting your "hideaway" into a "commercial enterprise." Some will advise you that you must have a lawyer. Others, like us, will be inclined to be slow about bringing one into the picture because we are convinced that the single most significant factor in a tough legal situation is power - raw power, not legal niceties or obeying the law or who is right or wrong.

Here are two specific examples to illustrate the point:

1. The case of a prominent U.S. Senator, able to have a party, takes a girl out afterward where she was killed, and never has to account for it the way an ordinary citizen would.

   That is the law. That is not right. That is raw power.

This was a criminal situation which was handled within the scope of the law and hushed up in a remarkably short time; but one can only wonder what might have happened if the victim had been a powerful person, and the driver of the death car had been some ordinary guy named Joe.

2. Let us now consider a civil situation by comparison where the only common denominator is the idea that we have an ordinary party on one side and a powerful party on the other. This one dragged in the courts for over seven years. This writer is knowledgeable since he was with the side of the ordinary party.

   We shall call the ordinary party Cid and the powerful party Big GW.

Cid bought a large gold concession in Bolivia from an individual for $600,000 and began to prove it up with exploratory/ production work to verify the data in a geology report.

One day shortly after their project was started, another party at the opposite end of the property began to do essentially the same thing. It developed that the other party was a representative of Big GW, a giant American company in a bald faced act of claim jumping, obviously thinking they could get away with it because they were so big and powerful - the old bully routine.

The small operator, Cid, brought the case up in Bolivian court and in U.S. District Court in New York City, and quickly learned that Big GW might be right and could rig the court in Bolivia with consummate ease.

Fortunately, the judge in the U.S. Court found undisputable reasons why he could exercise jurisdiction over the loud objections of Big GW, and after a long dragged out pre-trial investigation, the judge ordered both parties to submit to binding arbitration under an individual with appropriate, superior competence in such a specialized situation.

An internationally recognized expert in such things accepted the assignment, and after thorough review of the facts awarded in favor of Cid, and recommended that the court order payment by Big GW of triple damages on $110,000,000 for a total of over $300,000,000 to Cid.

At that point Big GW reneged on the binding arbitration arguing over a technicality, wherein the judge attempted to decide in favor of Cid, but was thwarted by another technicality after more than five years of delays, set-asides, and extensions of time for collection of evidence from Bolivia, where Big GW, often appearing with as many as fifteen lawyers, argued one nit-picking point at a time with thirty day delays each time until the exasperated judge did something which caused Big GW to demand a hearing before a higher court of three judges.

In the last year the higher court sent the case back to the original judge and, after several more months, a condition developed where Cid would have to have several hundred thousand dollars to continue the fight. There was extensive travel required to foreign countries by numerous parties involving new depositions ad infinitum which Cid simply could not afford. Cid's lawyer had been working on a pure contingency, feeling so sure he would ultimately win.

Cid was beaten by raw power; and finally the case was thrown out of court on a technicality, after seven years of impossible odds.

<u>That is the law</u>. <u>That is not right</u>. <u>That is raw power</u>.

Final note: In a bitter-sweet act of ultimate justice, the Government of Bolivia canceled the concession to all parties, and in a political move turned it over to local Indians who tried to recover the gold with ordinary farm tractors and hand effort - an impossible situation; and the claims have largely surrendered back to the jungle.

* * * * * * * * * * * *

We reiterate that in ordinary circumstances you do not need the law or a lawyer as long as you make things clear and operate in a properly sincere way. In heavy circumstances it does not matter how you behaved. The most powerful side will win or you will lose so much money and time that it will not have been worth it to fight.

Our legal system in America is in shambles and half the lawyers will admit it. There are several contributing reasons:

1. We have several times more lawyers per person in America than in any other place on earth, and they have to find something to do.

2. We have a system which, thanks to our ideas of justice, leans far over backward to protect the criminal or the perpetrator of the foul deed from oppressive domination. We have swung from debtor's prisons to easy bankruptcy, as an example. No one wants a debtor's prison, but we still have to maintain a system where a debt is respected just the same.

3. There are so many cases in the system that everyone has to stand in line even to make a date to be heard. Any court clerk will advise you that the average time for litigation today is upwards of two years almost anywhere in this country.

4. There are so many laws on the books and so many very smart people ready to "interpret" them that lawyers can go on and on and on and on as long as you are ready to pay them their fee - which today is rarely less than $100 per hour plus the expenses - which can grow at an alarming rate when travel and depositions and research time come into play. You must remember that you are paying your lawyer to learn his job if there is anything new and different about your situation.

5. When you subject yourself to the legal system you enter a den where you are in there for one time only (you hope) while the lawyers involved are there every day. They know each other, as a rule, and the good-old-boy notions of any closed society apply. You are no more than a pawn at lunch where the lawyer with the best trick in what they regard to be a simple adversarial game will win, hands down.

6. The judge himself is a lawyer.

7. There is no place for morality in the system where no one spends even a moment thinking about what is right and what is wrong. The lawyer learns as part of his basic education to disregard anything but the facts of the law - precedent and legislation - and the best lawyer is the one with the best bag of tricks.

8. One out of four lawyers and judges are thought to function while under the influence of drugs or alcohol in this country; and the California State Bar Association has publicly admitted this.

9. More and more lawyers are being brought up for malpractice, but you have to have a powerful, airtight case or no lawyer will attempt to cut down a member of his own fraternity. Few lawyers will take a malpractice case at all, as it is.

* * * * * * * *

So what are you supposed to do about all this when you decide to convert what was going to be a "hideaway" into an operating, income producing mine after you found that unexpected big bonanza?

<u>Rule 1</u> Do not try to play games.

<u>Rule 2</u> Make absolutely sure of all your facts. Do not assume anything, if it is in the least important.

<u>Rule 3</u> Do not attempt to do business if there is anything doubtful about the people involved. Check up on whom you are doing business with - always. Do not wait to find out critical details in a deposition someday.

<u>Rule 4</u> Make sure your deal is plain and simple to understand with no opportunity for misunderstanding.

<u>Rule 5</u> Do not count on any oral statements or understandings at all. Get it in writing that is clear and unencumbered with words that you do not understand.

<u>Rule 6</u> Do not try to nor expect to get something for nothing. When someone offers to do you a favor of any kind - don't take it,

<u>Rule 7</u> In dealing with any entity that is bigger or more powerful than you are, know that you have to be off on the right foot from the start. If things take a bad turn, remember that fine Kenny Rogers song:
"You have to know when to hold 'em, -
Know when to fold 'em, -
Know when to walk away -
And don't count your winnin's 'til the game is done."

The cornerstone of any activity among people is an agreement.

1. Today, the only kind of agreement that counts is in writing. Judges have a habit of disregarding anything that you can't put through a copy machine.

   There is another good, practical reason for this. When the parties see in writing what they think they have made clear in talk, it is not uncommon to discover a discrepancy; and careful study of the writing before you sign is a good insurance of clear understanding before it is too late.

2. All parties to an agreement must be legally capable of contracting. Generally, everyone is capable of contracting except parties who are limited in some way (too young, unsound mind, alien, deprived of civil rights [convicts], etc.)

3. There must be mutual consent. The parties must be free of pressure. There must be no menace, undue influence, threats, or duress. A fraud or a mistake can void a contract. Courts don't read minds.

4. There must be a specific offer and a specific acceptance or consent.

   There is a difference between an "offer" and an "invitation." A social invitation is not an offer. An advertisement is usually considered to be an invitation not an offer.

   The offer or acceptance must be free of illusion. You do not have a deal, for example, if the party making you an offer conditions it upon obtaining a loan. He may simply never try to get the loan, and you are left holding the bag. A general example to be avoided is the case where an offeror or offeree's promise is completely within his control to perform or not to perform.

5. Any agreement is useless unless the terms are clear, certain, and definite.

6. An agreement requires a consideration of time. When does it come to an end? How do you get off the train?

7. The contemplated activity under the contract must be lawful.

8. There must be a provision in the event of a disagreement or lack of understanding. Here, we say again that any agreement is only as good as the person agreeing. In the old days a handshake would do, and good

men were proud of it. Now, the handshake won't do, but among parties to a contract, the principle should still apply. The idea was that no contract could cover all contingencies as you go along, but the parties resolve to abide by the "spirit and intent."

We are learning to avoid the court and lawyers everywhere we can; and the old idea of having a will is about passé as we lean more and more on Living Trusts to avoid probate and the long delays. This can be especially important with respect to your mine claim.

Similarly, in contracts, we are choosing to arbitrate because it is more practical, time saving, cost saving, and equally as good as a legal finding if it is done right.

Here is an extract of a publication by the American Arbitration Association stating the Standard Arbitration Clause which may be used:

"Any controversy or claim arising out of or relating to this contract, or the breach thereof, shall be settled by arbitration in accordance with the rules of the American Arbitration Association, and judgement upon the award rendered by the Arbitrator(s) may be entered in any Court having jurisdiction thereof."

Finally, there is the point of paying for litigation, and we think it is fair to include the idea that the losing party in any dispute will pay all the legal bills. But, that is up to you.

"Then it's agreed. Watson, Smith, Teller, and Wilson go to Heaven; Jones, Paducci, and Horner go to Hell; and Fenton and Miller go to arbitration."

You might say to yourself that this is all so complicated you must hire a lawyer just to be sure you don't leave anything out from the very beginning; and there is some merit in this argument. But let us point out a few things:

1. Your lawyer cannot do your thinking for you - even though many of them will try. You are the person in charge, and the person ultimately responsible. When they clang the big steel door, your lawyer will be watching you from the outside.

2. Your lawyer will be looking out for himself first - before you. When you ask him to make sure "everything is covered," which is your primary concern early on, he will simply press a bunch of buttons in his computer and insert several pages of "boiler plate" paragraphs to cover the waterfront. And all this does is leave openings for his opponent when you come to the big court fight.

3. In any court, the judge has a legal obligation when you do the contract to try to figure out what the spirit and intent really was, and he knows he has to think in terms of what words ordinarily mean or what you really meant. He cannot fall back on anything else <u>unless</u> the wording is created by an attorney. Then you have no benefit of the judge's discretion. We have found that clearly written contracts covering the salient points as clearly as you can, after thinking a lot, are every bit as good and often better than windy documents from lawyer's offices where the lawyer had not given your problem more than cursory thought and leans on his manual more than trying to address your situation specifically.

Years ago we had the privilege of sharing a house on a bachelor basis with two great men. We three were operating regularly day-in and day-out in negotiations with oil companies in Canada as the new oil boom was exploding shortly after World War II. One minute we were dealing with the giant Hudson Bay Oil & Gas Company, and in the next we were helping some new promotion before the ink was dry on their permits. This writer was providing technical expertise as a licensed geologist and geophysicist and a Registered Professional Engineer. One of the men we refer to was a professional manager, one time head man at the Rand McNally Map Co. and then Vice President of General Motors in charge of their radiator division. He was full of jokes that always had a point to them. Many a truth is spoken in jest.

The other man, who we shall call HBB, was a lawyer and secretary to the Twelve Apostles of the Church of Jesus Christ of Latter-day Saints, the Mormons. He had very little use for attorneys in general and made no effort to hide it. He was a true believer in the principle of the law and the idea of justice, but he sputtered constantly about the application in day-to-day court activity.

One evening he said, "You know there is no substitute for keeping it short and sweet. We could be sitting here on this porch. I could be eating an orange and say to you, 'would you like one?' You might say yes and I would simply give it to you. That's that.

"Now if we did that by a contract and I said I, HBB, of such and such a place, the owner of that certain orange, etc., etc., do hereby give, bequeath, and devise the orange to you, WT, of such a place. Then we would think of all the things you might do with it - to eat, suck on, squeeze, make juice, marmalade, cake, etc., etc. Then, there would be a place on the paper to accept this according to the terms and conditions set out herewith, etc. etc. We date it and sign it. Then I go away.

"Five years later I come back to see you and we're sitting right here on this same porch. I look around and say, 'That sure is a fine young tree you have there.' You say, 'Remember that orange? Well, I dried the seeds and planted them and that tree is what came up.'

"Well, I say, I don't recollect saying anything in that contract about planting trees! I do believe that tree is mine."

We all laughed, and then fell silent, just thinking for a while.

NOTICE: The information in this chapter has been presented as opinion of the writer, and not legal advice.

The writer is not an attorney.

10. *FAMILY/FRIENDS and ODDS and ENDS*

Having touched on the remote possibility that you might discover a big deposit of gold or some other valuable mineral after hundreds of smart, tough old timers have passed over the ground, we shall hope you have found a pleasant spot, as the old song goes, "Somewhere Out In The West."

We make a joke, you think? Not so. We still say it is a very real possibility; and you can enjoy it as a place to go to year after year for a long time.

To be practical, if you are on some kind of stream where a dredge will work, there is no reason why you can not rent the place by the week or something like that to friends or people you can trust, when you are not using it. On twenty acres you could have three or four fine campsites which could be occupied simultaneously.

Your friends can camp out and look for a little gold. It need not necessarily be a dredge situation. It might be a seam in a hill where you could follow a quartz vein and find beautiful specimens laced with gold that are very attractive in many applications. Be careful about going underground, because that can be very dangerous.

You know the claim you located must be a place where you are truly prospecting at least some of the time, but there is no law that says you can't go fishing from there or hunting if you are in the right location. It can be used as a base for hiking or prospecting in other areas nearby, and with the continuously developing, improved devices for finding precious metals, you certainly could find things an old timer might well have passed over.

The early old timers in the West, men wild with hope and dreams, seemed to splash over the countryside motivated by rumors and tall tales. If they didn't find a pound of gold a day practically lying on the ground, they hustled on to the next place. These were the people, shall we say, from 1840-1860 who were chasing madly after gold.

Between 1860 and 1900 the old timers became much more systematic. Many of them were also knowledgeable, having heard about the findings, and came here from Europe or China as experienced miners who knew how to timber tunnels, read rivers, track out ore veins, and find the very best spots for accumulation of minerals. They were no fools, and most of us who have come in behind them have learned to have a great deal of respect for their prowess.

Living in populated places with automatic controls of heat and air around us all the time, stoves we can turn on with a button, water from taps, and all the rest, it is difficult to imagine that any of us could recreate in ourselves the personal capabilities or resourcefulness of those people.

There is a lot of romance in providing for your own food by going into the wilderness and hunting or fishing, but we are so far from those days it is impossible to imagine us leaving the freeway and even staying alive any longer than it takes food to spoil.

MAJOR WILLIAM DOWNIE
One of the greatest old-timers of them all.

With our computers and outer space programs and cruise missiles and smart bombs and microwave ovens and television sets and credit cards, we are often prone to think people before us barely knew enough to get along, but with a few years in the wilderness exposed to the deeds of the people who have trod the mountains before us, we gain a whole new perspective.

We could fill a small book with information we have stumbled on over the years as we turned up tidbits of unusual words, ideas, and abilities common to many of these old guys that are now almost all gone. For example, there were many words related to the moon that were readily used and understood in those days that we never hear any more. Just think about the name of this writer's mine - Deep Moon. What does that mean? Deep Moon? You tell us. How about the words, deep space, a relatively new scientific term that probably would not have meant anything to the man who knew what a deep moon was? We had to learn a new vocabulary dealing with the solar system to know what deep moon means.

You say new moon, old moon, full moon, etc. Full is when it looks the roundest and is opposite the sun. If you think carefully about the geometry, you would see this kind of moon from sunset to the following sunrise, and the moon is "high" at midnight. You see it follows that it is "deep" when it is opposite that point, actually on the other side of the earth, or in a sense, "deepest" through the earth yet still full, which would actually be at "high noon." If you are not thoroughly confused by now, perhaps we should try again! Good luck.

Our units of measure are all precisely defined today by meter bars made out of illustrious metals or by some kind of optical measurement tied to computer chips in one way or another, but there was a day when a foot meant the length of the King's foot. We still use the word, hand (equal to four inches) in measuring the height of horses. In the old days this was the width of your hand when held flat with your thumb against your first finger. Try measuring yours, if you are not too big or too small, you will see that it is very close. The word inch was once defined as the width of the King's thumb and, because it was about one twelfth of the length of his foot, we have from Old English, ynce. This is related to the word ounce, but we can not get too far off course here in a book about gold and hideaways.

We will call your attention to one other old concept, very useful in estimating distances and heights.

The average person's eyes are almost exactly as far above his feet as his stride is long. A stride, as you probably know, is two steps, the distance between the point where your foot comes down when you are moving along and where the same foot comes down again.

You will find that if you walk naturally, your stride will be just about five feet and twenty strides would be a hundred feet, and so on. It is very worthwhile to evaluate your stride over a measured distance. Be sure to walk very naturally. Some people, stepping off a distance, will step as far as they can, really stretching out, and in this case their stride might be as much as six feet. This may be all right on easy flat ground, but if you are moving over any irregular surface it is much better to rely on the ordinary walking dimension. If you check yourself and know what your stride length is, you will be amazed at how accurately you can step off a distance of as much as a quarter of a mile. Remember your stride will shorten some if you are moving up or down hill.

If you are medium height, your eyes will be very close to five feet off the ground. If you sight something at a bit of a distance and visualize the horizon behind the object, you can spot a height that is very close to five feet above your feet. This is very useful in estimating differences in elevation.

From a purely intellectual point of view we could wonder why we all do not simply use the metric system in all ways; but the old ideas are so deeply imbedded in our culture that it will probably be a long time before we change completely, if we ever do.

For example, when we were thinking about placer claim locations and the township and range system in this country, we were talking about miles, 5,280 feet. Why such an odd number? Well it goes back to all those old ideas we have hinted when a yard and a rod and a chain were routinely used as units of measure on the ground. A yard is three feet, as you know. Why three? Why not five or ten, or something at least logical. A rod is 16-1/2 feet or 5-1/2 yards. What is going on? How weird you might say. But now we have a chain, which is 66 feet, 4 rods, composed of 100 links of 8 inches (2 hands) each.

That 66 foot distance comes to us today in the township and range system in such a way that a standard placer claim, instead of being 600 by 1500 feet, as we indicated a lode claim would be in Chapter 4, is 660 by 1320 feet, or ten by twenty Gunter's chains in size.

Why do we belabor these odd problems in land measurement? Simple. The moment you begin to look around in the west for that placer claim, you will run straight into all these complications and you may as well be prepared. Don't be discouraged or alarmed. It is really not too difficult, and you will get used to it.

Once you become involved you will discover that there is a great deal of fun in looking for a nice gold claim. There is a great deal of fascinating information to be studied and thought through.

On a step by step basis, to sum things up, we think you should probably acquire a few books and maps covering the whole country from New Mexico, Colorado, Wyoming, and Montana west to the Pacific Ocean. Remember too, that there are spots from northern Virginia to central Alabama that could prove interesting.

Next, delineate the area where you want to look in a general way, and find out where the federal land is in that part of the country. You can forget the National Parks except for claims that were recorded on or before September 28, 1977. You may assume that anyone who owns one of those today probably has a very high price tag on it, if at all. There are a great many people who would dearly like to have you out of a claim like that, so think about it carefully as the years pass and imagine yourself trying to enjoy life mining in the sights of all those public minded types who will be coming down on you harder and harder each year. The only safe claim would be a patented claim, if you can find it.

Military Reservations, Weapons Centers, Military Bases, Firing and Bombing Ranges take up hundreds of thousands of acres in the west and are clearly off limits to the kind of activity we contemplate here.

Forget about Indian Reservations, State owned land, National Monuments like Death Valley (a bit hot, you know), Wilderness Areas or lands which have been withdrawn or classified against mining location, or as "required lands." But, not to worry! There is still a whole lot of country left to play around in.

Having chosen the broad area, the next thing to do is search out the ordinary roads and towns and rivers, and then divide the area up into small, ten mile stretches or so, and give those areas a priority for further work.

Now, armed with some good, detailed maps of "priority areas," pack your lunch or whatever, take your binoculars and your note pad, your compass and camera, and have at it. You will find beautiful spots on or near the rivers or up the mountainsides. Keep your eye out for claim markers. They are hard to spot at first but knowledgeable folks soon learn to spot them. You develop a sense for that kind of thing after a bit of experience.

Make precise notes of the location of favorite areas, being sure you can identify them in terms of the Township-Range System we talked about in Chapters 2, 3 and 4.

Now check your finds at the BLM office, and make copies of the films that apply, showing you geographic locations, names and addresses of claim owners and names of claims, being sure to define if they are placer or lode claims and if the assessment work is up to date. While you are at it, you might consider getting the word about your "priority area 2" or even more country. It could save you a trip later.

When you find an area that really looks good, it is time to visit the County Recorder's Office and have a look at the records. You can look things up by claim name, but it is easier most of the time to look up the records by the name of the claim owner. Watch for any recordings of contracts or other obligations against the claim you like, and make copious notes.

Once you get that far, you won't need us any more. You will be an expert yourself, ready to trade with some guy young or old who is either ready to talk to you or ready to throw you off the property.

You can find that fine claim, and find some gold too. It really can be a hideaway you can enjoy for years, make money on if you are smart, and then - just watch it grow in value as the years go by. We have such conditions as "grandfather status"; and you could find yourself "grandfathered" into a situation that could not be duplicated a few years from now. Think about that!

BIBLIOGRAPHY

Floating Dredges: California Geology, June 1972 Issue, California Division of Mines and Geology, P.O. Box 2980, Sacramento, California 95812

Law: Digest of Mining Claim Laws, Robert G. Pruitt, Jr., Rocky Mountain Law Foundation, Fleming Law Building, University of Colorado, Boulder, Colorado

Mining Law, from Location to Patent, Terry S. Maley, Mineral Land Publications, P.O. Box 1186, Boise, Idaho 83701

The actual text of laws can be found in the Federal Register, published daily except Saturday, Sunday, and Monday and days after holidays in Washington, DC.

For all practical purposes circulars and reprints of regulations are available from most state offices of the BLM, Bureau of Land Management.

Maps: National Cartographic Information Center, U.S. Department of the Interior, Geological Survey, Reston, Virginia 22092 This is the basic reference. You can get published maps, topographic maps, orthophotoquads, state maps, vicinity maps, satellite image maps, United States maps and world maps from their sub-centers at:

Western Mapping Center - NCIC, U.S. Geological Survey, 345 Middlefield Road, Menlo Park, CA 94025

Western Distribution Branch, U.S. Geological Survey, Building 41, Box 25286, Federal Center, Denver, Colorado 80225

For keeping up with the world of mine claims and mining we suggest you read the CALIFORNIA MINING JOURNAL, a monthly publication covering mining around the world in a very authoritative way. Their address is: P.O. Box 2260, 9032 Soquel Drive, Aptos, CA 95001, telephone (408) 662-2899.

General

California Gold Camps
Erwin G. Gudde, University of California Press, 1975 - an invaluable reference to the old camps, towns and localities with geographical and historical notes. Unique illustrations.

California Place Names
Erwin G. Gudde, University of California Press, 1969 - a clear, authoritative, in-depth analysis of the old words from Aztec, Miwok, Spanish, and other sources.

Fool's Gold
Richard Dillon, Western Tanager Press, 1967, the story of John Augustus Sutter, "The Father of California."

Forty-Niners
Archer Butler Hulbert, Little, Brown and Company, 1949, "a complete and connected story of what happened to those thousands of men and women who toiled through the dust and misery of the Overland Trail."

Gold Districts of California
William B. Clark, Bulletin 193, Calif. Div. of Mines and Geology, Fifth Printing 1980, an official, overall guide to the gold deposits in California. Copiously illustrated.

Hunting For Gold
Major William Downie, American West Publishing Co., first published in 1893, republished in 1971. This is the most respected first hand account of the 49er period in the history of the West from California to Alaska.

Mineral Resources of California
Bulletin 191, Calif. Div. of Mines and Geology, 1966, the official technical book.

Realms of Gold
Ray Vicker, Charles Scribner's Sons, 1975 "covers every aspect of the long story of gold from its ceremonial role to its monetary use."

You can find equipment in sporting goods and camping and hiking stores all over the country. Many of them also carry the U.S.G.S. topographic maps, but there are a few sources you just might not know:

<u>Miners Catalog</u>	P.O. Box 1301, Riggins, Idaho 83549, telephone (208) 628-3247. They cater to explorers, miners, and surveyors. They also carry a number of books, listed in their catalog.
<u>Northern</u>	P.O. Box 1499 Burnsville, Minnesota 55337, telephone (800) 533-5545. They carry a broad spectrum of tools and equipment with excellent service and good prices.
<u>Keene Engineering</u>	9380 Corbin Avenue, Northridge, California 91324. They claim to be the world's largest manufacturer of portable mining equipment.

The ILLUSTRATIONS

All illustrations provided by the author except as follows:

page 20 and 108 from *Hunting for Gold* by William Downie
The California Publishing Company 1893

page 32 from *A Manual of Topographic Methods*, Henry Gannett,
U.S. Geological Survey Monograph Vol. XXII,
Washington, D.C. 1893

page 39 dredging on the Yuba River from Tahoe National Forest Map

page 42 old drawing of Downieville from Bancroft Library,
University of California, Berkeley, CA

page 56 from *Diving For Gold*, William B. Clark, California Geology,
April 1972, California Division of Mines and Geology

page 61-63 from California Mining Journal

page 74 from *The Indian Tipi*, Reginald & Gladys Laubin,
University of Oklahoma Press, 1957

page 94 from old photo - source unknown

INDEX

About the AUTHOR

Born in a very small New Hampshire town, the author commenced his education in a one room school, going on to Andover, Harvard, and Norwich Military Academy, earning a degree in English Literature and American History and a commission in the U.S. Cavalry.

He went on to do graduate work in geology, geophysics, geography, economics, and business administration at Columbia University, University of Michigan, University of Southern California, and U.C.L.A.

Functioning as a line officer of the U.S. Cavalry he was trained, and operated with the 2nd Corps Area Headquarters, in G-2, Military Intelligence, and was subsequently assigned to West Point as a regular instructor. He resigned from the military in protest of its political use in Korea, and went into commercial exploration with the Standard Oil Company of New York (SOCONY).

Later, accepted in Canada as a Registered Professional Engineer, he was Chief Geophysicist for Pacific Petroleums, Ltd. (listed today on the New York Stock Exchange) and Bear Oil Company (since disbanded), a large consortium of American oil companies exploring in the far north. He subsequently accepted the position of Executive Vice President (CEO) of Overland Industries Ltd. (since sold out to Imperial Oil, Hudson Bay Oil & Gas, and others), while operating his own exploration company, Tellington Associates of Canada, Ltd.

Over the years he has been involved in extractive mineral mining and exploration in Canada, the United States, and Central and South America. He is the author of several books and the inventor of "The Vehicle Tracker," grandfather device behind today's guidance systems used to guide, track and report the location of vehicles in outer space. He was named by Governor Ronald Reagan for the Creative Citizenship Award in California in 1968. At the time of this writing, he owns a gold mining operation near Downieville, California.